PIEZOELECTRIC MATERIALS AND DEVICES- PRACTICE AND APPLICATIONS

Edited by **Farzad Ebrahimi**

Piezoelectric Materials and Devices- Practice and Applications
http://dx.doi.org/10.5772/45936
Edited by Farzad Ebrahimi

Contributors
Toshio Ogawa, Vahid Mohammadi, Saeideh Mohammadi, Fereshteh Barghi, Saeed Assarzadeh, Sebastien Grondel, Christophe Delebarre, Andrzej Buchacz, Andrzej Wróbel, Sonia Djili, Farouk Benmeddour, Emmanuel Moulin, Jamal Assaad, Farzad Ebrahimi

Published by InTech
Janeza Trdine 9, 51000 Rijeka, Croatia

Publishing Process Manager Danijela Duric
Technical Editor InTech DTP team
Cover InTech Design team

Additional hard copies can be obtained from orders@intechopen.com

Piezoelectric Materials and Devices- Practice and Applications, Edited by Farzad Ebrahimi
p. cm.
ISBN 978-953-51-1045-3

For EU product safety concerns:
IN TECH d.o.o., Prolaz Marije Krucifikse Kozulić 3, 51000 Rijeka, Croatia,
info@intechopen.com or visit our website at intechopen.com.

Meet the editor

Dr. Farzad Ebrahimi was born in Qazvin, Iran, in 1979. He graduated in mechanical engineering, from the University of Tehran, Iran, in 2002. He received his Msc and PhD in mechanical engineering, with a specialization in applied design from the University of Tehran, Iran, in 2009. Since 2002, he has been working at the "Smart Materials and Structures Lab" Research Center of the faculty of mechanical engineering at the University of Tehran, where he is a researcher of smart functionally graded materials and structures. He began his university carrier as an assistant professor in the department of mechanical engineering at Imam Khomeini International University, Qazvin.

Contents

Preface

Piezoelectricity from the Greek word "piezo" means pressure electricity. Certain crystalline substances generate electric charges under mechanical stress and conversely experience a mechanical strain in the presence of an electric field.

The piezoelectric effect was discovered in some naturally occurring materials in the 1880s. However it was not until the Second World War that man-made polycrystalline ceramic materials were produced that also showed piezoelectric properties. Quartz and other natural crystals found application in microphones, accelerometers and ultrasonic transducers, whilst the advent of the man-made piezoelectric materials widened the field of applications to include sonar, hydrophones, and piezo-ignition systems.

Piezoelectric materials are widely used to make various devices including transducers for converting electrical energy to mechanical energy or vice-versa, sensors, actuators, and resonators and filters for telecommunication, control and time-keeping. Piezoelectric materials also act an important role in healthcare applications such as echographic images, new imaging techniques, and use in ultrasonic surgery, miniature sensors and hearing aids and in the transportation industry such as sensor performance devices, actuators, air bag sensors, micro-pumps and micro-motors.

Since its discovery, the piezoelectricity effect has found many useful applications, such as the production and detection of sound, generation of high voltages and frequency, microbalances, and ultra fine focusing of optical assemblies. It is also the basis of a number of scientific instrumental techniques with atomic resolution, the scanning probe microscopies such as STM, AFM, MTA, SNOM, etc., and everyday uses such as acting as the ignition source for cigarette lighters and push-start propane barbecues.

Piezoelectric devices are a very reliable and inexpensive means of converting electrical energy into physical motion and exhibit a high tolerance to environmental factors such as electromagnetic fields and humidity.

Today, piezoelectric applications include smart materials for vibration control, aerospace and astronautical applications of flexible surfaces and structures and novel applications for vibration reduction in sports equipment. Piezoelectric devices are integral components of common devices such as printers, stereos, electric guitars and gas igniting lighters used by millions on a daily basis. The piezoelectric ceramic transducers have also found many important applications in adaptive structures for vibration control and acoustic noise suppression in modern space, civilian and military systems, such as launch vehicles, space platforms, aircraft, submarines and helicopters.

This book reports on the state of the art research and development findings on important issues at the field of **piezoelectric materials and devices-practice and application** such as piezoelectric stacks in level sensors, pressure sensors and actuators for functionally graded plates. It also presents related themes in the field of application of piezoelectric devices in active and passive health monitoring systems, machining processes, nondestructive testing of aeronautical structures and acoustic wave velocity measurement.

The text is addressed not only to researchers, but also to professional engineers, students and other experts in a variety of disciplines, both academic and industrial seeking to gain a better understanding of what has been done in the field recently, and what kind of open problems are in this area.

I hope that readers will find the book useful and inspiring by examining the recent developments in **piezoelectric materials and devices** .

Dr. Farzad Ebrahimi
Faculty of Engineering,
Mechanical Engineering Department,
International University of Imam Khomeini
Qazvin, I. R. IRAN

Chapter 1

Piezoelectric Actuators for Functionally Graded Plates-Nonlinear Vibration Analysis

Farzad Ebrahimi

Additional information is available at the end of the chapter

http://dx.doi.org/10.5772/52407

1. Introduction

Functionally graded materials (FGMs) are a new generation of composite materials wherein the material properties vary continuously to yield a predetermined composition profile. These materials have been introduced to benefit from the ideal performance of its constituents, e.g., high heat/corrosion resistance of ceramics on one side, and large mechanical strength and toughness of metals on the other side. FGMs have no interfaces and are hence advantageous over conventional laminated composites. FGMs also permit tailoring of material composition to optimize a desired characteristic such as minimizing the maximum deflection for a given load and boundary conditions, or maximizing the first frequency of free vibration, or minimizing the maximum principal tensile stress. As a result, FGMs have gained potential applications in a wide variety of engineering components or systems, including armor plating, heat engine components and human implants. FGMs are now developed for general use as structural components and especially to operate in environments with extremely high temperatures. Low thermal conductivity, low coefficient of thermal expansion and core ductility have enabled the FGM materials to withstand higher temperature gradients for a given heat flux. Structures made of FGMs are often susceptible to failure from large deflections, or excessive stresses that are induced by large temperature gradients and/or mechanical loads. It is therefore of prime importance to account for the geometrically nonlinear deformation as well as the thermal environment effect to ensure more accurate and reliable structural analysis and design.

The concept of developing smart structures has been extensively used for active control of flexible structures during the past decade [1-3]. In this regard, the use of axisymmetric piezoelectric actuators in the form of a disc or ring to produce motion in a circular or annular substrate plate is common in a wide range of applications including micro-pumps and micro-valves [4, 5], devices for generating and detecting sound [6] and implantable medical devices [7]. They may also be useful in other applications such as microwave micro-switches where it is important to control distortion due to intrinsic stresses [8]. Also in recent years, with the increasing use of smart material in vibration control of plate structures, the mechanical response of FGM plates with surface-bonded piezoelectric layers has attracted some researchers' attention. Since this area is relatively new, published literature on the free and forced vibration of FGM plates is limited and most are focused on the cases of the linear problem. Among those, a 3-D solution for rectangular FG plates coupled with a piezoelectric actuator layer was proposed by Reddy and Cheng [9] using transfer matrix and asymptotic expansion techniques. Wang and Noda [10] analyzed a smart FG composite structure composed of a layer of metal, a layer of piezoelectric and an FG layer in between, while He et al. [11] developed a finite element model for studying the shape and vibration control of FG plates integrated with piezoelectric sensors and actuators. Yang et al. [12] investigated the nonlinear thermo-electro-mechanical bending response of FG rectangular plates that are covered with monolithic piezoelectric actuator layers on the top and bottom surfaces of the plate. More recently, Huang and Shen [13] investigated the dynamics of an FG plate coupled with two monolithic piezoelectric layers at its top and bottom surfaces undergoing nonlinear vibrations in thermal environments. In addition, finite element piezothermoelasticity analysis and the active control of FGM plates with integrated piezoelectric sensors and actuators was studied by Liew et al. [14] and the temperature response of FGMs using a nonlinear finite element method was studied by Zhai et al. [15]. All the aforementioned studies focused on the rectangular-shaped plate structures. To the authors' best knowledge, no researches dealing with the nonlinear vibration characteristics of the circular functionally graded plate integrated with the piezoelectric layers have been reported in the literature except the author's recent works in presenting an analytical solution for the free axisymmetric linear vibration of piezoelectric coupled circular and annular FGM plates [16-20] and investigating the applied control voltage effect on piezoelectrically actuated nonlinear FG circular plate [21] in which the thermal environment effects are not taken in to account.

Consequently, a non-linear dynamics and vibration analysis is conducted on pre-stressed piezo-actuated FG circular plates in thermal environment. Nonlinear governing equations of motion are derived based on Kirchhoff's-Love hypothesis with von-Karman type geometrical large nonlinear deformations. Dynamic equations and boundary conditions including thermal, elastic and piezoelectric couplings are formulated and solutions are derived. An exact series expansion method combined with perturbation approach is used to model the non-linear thermo-electro-mechanical vibration behavior of the structure. Numerical results for FG plates with various mixture of ceramic and metal are presented in dimensionless forms. A parametric study is also undertaken to highlight the effects of the thermal environment, applied actuator voltage and material composition of FG core plate on the nonlinear vibration characteristics of the composite structure. The new features of the effect of thermal environment and applied actuator voltage on free vibration of FG plates and some meaningful results in this chapter are helpful for the application and the design of nuclear reactors, space planes and chemical plants, in which functionally graded plates act as basic elements.

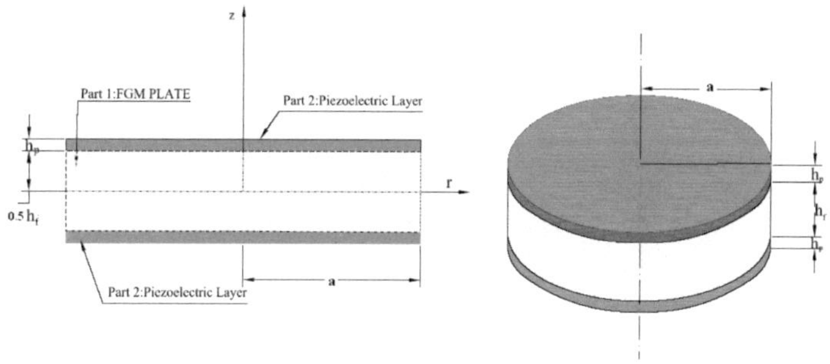

Figure 1. FG circular plate with two piezoelectric actuators.

2. Functionally graded materials (FGM)

Nowadays, not only can FGM easily be produced but one can control even the variation of the FG constituents in a specific way. For example in an FG material made of ceramic and metal mixture, we have:

$$V_m + V_c = 1 \tag{1}$$

in which V_c and V_m are the volume fraction of the ceramic and metallic part, respectively. Based on the power law distribution [22], the variation of V_c vs. thickness coordinate *(z)* with its origin placed at the middle of thickness, can be expressed as:

$$V_c = (z/h_f + 1/2)^n \quad ,n \geq 0 \tag{2}$$

in which h_f is the FG core plate thickness and n is the FGM volume fraction index (see Figure 1). Note that the variation of both constituents (ceramics and metal) is linear when $n=1$. We assume that the inhomogeneous material properties, such as the modulus of elasticity E, densityρ, thermal expansion coefficient α and the thermal conductivity κ change within the thickness direction z based on Voigt's rule over the whole range of the volume fraction [23] while Poisson's ratio υ is assumed to be constant in the thickness direction [24] as:

$$\begin{aligned}
&E(z) = (E_c - E_m)V_c(z) + E_m, \\
&\rho(z) = (\rho_c - \rho_m)V_c(z) + \rho_m \\
&\alpha(z) = (\alpha_c - \alpha_m)V_c(z) + \alpha_m, \\
&\nu(z) = \nu \\
&\kappa(z) = (\kappa_c - \kappa_m)V_c(z) + \kappa_m
\end{aligned} \tag{3}$$

where subscripts m and c refer to the metal and ceramic constituents, respectively. After substituting Vc from Eq. (2) into Eqs. (3), material properties of the FGM plate are determined in the power law form-the same as those proposed by Reddy and Praveen [22]:

$$
\begin{aligned}
E_f(z) &= (E_c - E_m)(z/h_f + 1/2)^n + E_m, \\
\rho_f(z) &= (\rho_c - \rho_m)(z/h_f + 1/2)^n + \rho_m, \\
\kappa_f(z) &= (\kappa_c - \kappa_m)(z/h_f + 1/2)^n + \kappa_m, \\
\alpha_f(z) &= (\alpha_c - \alpha_m)(z/h_f + 1/2)^n + \alpha_m
\end{aligned}
\tag{4}
$$

3. Thermal environment

Assume a piezo-laminated FGM plate is subjected to a thermal environment and the temperature variation occurs in the thickness direction and 1D temperature field is assumed to be constant in the r-θ plane of the plate. In such a case, the temperature distribution along the thickness can be obtained by solving a steady-state heat transfer equation

$$
-\frac{d}{dz}\left[\kappa(z)\frac{dT}{dz}\right] = 0 \tag{5}
$$

in which

$$
\kappa(z) = \begin{cases} \kappa_p & (h_f/2 < z < h_p + h_f/2) \\ \kappa_f(z) & (-h_f/2 < z < h_f/2) \\ \kappa_p & (-h_p - h_f/2 < z < -h_f/2) \end{cases} \tag{6}
$$

$$
\kappa(z) = \begin{cases} \kappa_p & (h_f/2 < z < h_p + h_f/2) \\ \kappa_f(z) & (-h_f/2 < z < h_f/2) \\ \kappa_p & (-h_p - h_f/2 < z < -h_f/2) \end{cases} \tag{7}
$$

where κ_p and κ_f are the thermal conductivity of piezoelectric layers and FG plate, respectively. Eq. (5) is solved by imposing the boundary conditions as

$$\begin{aligned} T_p\big|_{z=h_p+h_f/2} &= T_U \\ \tilde{T}_p\big|_{z=-h_p-h_f/2} &= T_L \end{aligned} \tag{8}$$

and the continuity conditions

$$\begin{aligned} &T_p\big|_{z=h_f/2} = T_f\big|_{z=h_f/2} = T_1, \\ &T_f\big|_{z=-h_f/2} = \tilde{T}_p\big|_{z=-h_f/2} = T_2, \\ &\kappa_p \frac{dT_p(z)}{dz}\bigg|_{z=h_f/2} = \kappa_c \frac{dT_f(z)}{dz}\bigg|_{z=h_f/2}, \\ &\kappa_p \frac{d\tilde{T}_p(z)}{dz}\bigg|_{z=-h_f/2} = \kappa_m \frac{dT_f(z)}{dz}\bigg|_{z=-h_f/2} \end{aligned} \tag{9}$$

The solution of Eq.(5) with the aforementioned conditions can be expressed as polynomial series:

$$T_p(z) = T_1 + \frac{T_U - T_1}{h_p}\left(z - h_f/2\right) \tag{10}$$

$$\tilde{T}_p(z) = T_L + \frac{T_2 - T_L}{h_p}\left(z + h_f/2 + h_p\right) \tag{11}$$

and

$$\begin{aligned} T_f(z) = {} & A_0 + A_1\left(\frac{z}{h_f}+\frac{1}{2}\right) + A_2\left(\frac{z}{h_f}+\frac{1}{2}\right)^{n+1} + A_3\left(\frac{z}{h_f}+\frac{1}{2}\right)^{2n+1} + A_4\left(\frac{z}{h_f}+\frac{1}{2}\right)^{3n+1} + \\ & A_5\left(\frac{z}{h_f}+\frac{1}{2}\right)^{4n+1} + A_6\left(\frac{z}{h_f}+\frac{1}{2}\right)^{5n+1} + O(z)^{6n+1} \end{aligned} \tag{12}$$

where constants T_1, T_2 and A_j can be found in Appendix A.

4. Nonlinear piezo-thermo-electric coupled FG circular plate system

It is assumed that an FGM circular plate is sandwiched between two thin piezoelectric layers which are sensitive in both circumferential and radial directions as shown in Figure 1 and the structure is in thermal environment; also, the piezoelectric layers are much thinner than the FGM plate, i.e., $h_p \ll h_f$.An initial large deformation exceeding the linear range is imposed on the circular plate and a von-Karman type nonlinear deformation is adopted in the analysis. The von-Karman type nonlinearity assumes that the transverse nonlinear deflection w is much more prominent than the other two inplane deflections.

4.1. Nonlinear strain-displacement relations

Based on the Kirchhoff-Love assumptions, the strain components at distance z from the middle plane are given by

$$\begin{aligned} \varepsilon_{rr} &= \bar{\varepsilon}_{rr} + zk_{rr}, \\ \varepsilon_{\theta\theta} &= \bar{\varepsilon}_{\theta\theta} + zk_{\theta\theta}, \\ \varepsilon_{r\theta} &= \bar{\varepsilon}_{r\theta} + zk_{r\theta} \end{aligned} \tag{13}$$

where the z-axis is assumed positive outward. Here$\bar{\varepsilon}_{rr}$,$\bar{\varepsilon}_{\theta\theta}$, $\bar{\varepsilon}_{r\theta}$are the engineering strain components in the median surface, andk_{rr},$k_{\theta\theta}$, $k_{r\theta}$are the curvatures which can be expressed in terms of the displacement components. The relations between the middle plane strains and the displacement components according to the von-Karman type nonlinear deformation and Sander's assumptions [25] are defined as:

$$\begin{aligned} \bar{\varepsilon}_{rr} &= \frac{\partial u_r}{\partial r} + \frac{1}{2}\left(\frac{\partial w}{\partial r}\right)^2, \\ \bar{\varepsilon}_{\theta\theta} &= \frac{1}{r}\frac{\partial u_\theta}{\partial \theta} + \frac{u_r}{r} + \frac{1}{2}\left(\frac{1}{r}\frac{\partial w}{\partial \theta}\right)^2, \\ \bar{\varepsilon}_{r\theta} &= \frac{1}{r}\frac{\partial u_r}{\partial \theta} + \frac{\partial u_\theta}{\partial r} - \frac{u_\theta}{r} + \left(\frac{1}{r}\frac{\partial w}{\partial r}\right)\frac{\partial w}{\partial \theta} \end{aligned} \tag{14}$$

$$\begin{aligned} \kappa_{rr} &= -\frac{\partial^2 w}{\partial r^2}, \\ \kappa_{\theta\theta} &= -\frac{1}{r}\frac{\partial w}{\partial r} - \frac{1}{r^2}\frac{\partial^2 w}{\partial \theta^2}, \\ \kappa_{r\theta} &= -\frac{1}{r}\left(\frac{\partial^2 w}{\partial r \partial \theta}\right) + \frac{1}{2r^2}\frac{\partial w}{\partial \theta} \end{aligned} \tag{15}$$

where u_r, u_θ, w represent the corresponding components of the displacement of a point on the middle plate surface. Substituting Eqs. (14) and (15) into Eqs. (13), the following expressions for the strain components are obtained

$$\begin{aligned}
\varepsilon_{rr} &= \frac{\partial u_r}{\partial r} + \frac{1}{2}\left(\frac{\partial w}{\partial r}\right)^2 - z\frac{\partial^2 w}{\partial r^2}, \\
\varepsilon_{\theta\theta} &= \frac{1}{r}\frac{\partial u_\theta}{\partial \theta} + \frac{u_r}{r} + \frac{1}{2}\left(\frac{1}{r}\frac{\partial w}{\partial \theta}\right)^2 - z\left(\frac{1}{r}\frac{\partial w}{\partial r} + \frac{1}{r^2}\frac{\partial^2 w}{\partial \theta^2}\right), \\
\varepsilon_{r\theta} &= \frac{1}{r}\frac{\partial u_r}{\partial \theta} + \frac{\partial u_\theta}{\partial r} - \frac{u_\theta}{r} + \left(\frac{1}{r}\frac{\partial w}{\partial r}\right)\frac{\partial w}{\partial \theta} + 2z\left(-\frac{1}{r}\left(\frac{\partial^2 w}{\partial r\partial \theta}\right) + \frac{1}{2r^2}\frac{\partial w}{\partial \theta}\right)
\end{aligned} \tag{16}$$

For a circular plate with axisymmetric oscillations, the strain expressions are simplified to

$$\begin{aligned}
\varepsilon_{rr} &= \frac{\partial u_r}{\partial r} + \frac{1}{2}\left(\frac{\partial w}{\partial r}\right)^2 - z\frac{\partial^2 w}{\partial r^2}, \\
\varepsilon_{\theta\theta} &= \frac{u_r}{r} - \frac{z}{r}\frac{\partial w}{\partial r}, \\
\varepsilon_z &= \gamma_{r\theta} = \gamma_{\theta z} = \gamma_{zr} = 0
\end{aligned} \tag{17}$$

4.2. Force and moment resultants

The stress components in the FG core plate in terms of strains based on the generalized Hooke's Law using the plate theory approximation of $\sigma_z \approx 0$in the constitutive equations are defined as [26];

$$\sigma_r^f = \frac{E(z)}{1-\nu^2}(\varepsilon_r + \nu\varepsilon_\theta) - \frac{E(z)\alpha(z)}{1-\nu}\Delta T \tag{18}$$

$$\sigma_\theta^f = \frac{E(z)}{1-\nu^2}(\varepsilon_\theta + \nu\varepsilon_r) - \frac{E(z)\alpha(z)}{1-\nu}\Delta T \tag{19}$$

where E(z), ν(z) and $\alpha(z)$ are Young's modulus, Poisson's ratio and coefficient of thermal expansion of the FGM material, respectively, as expressed in Eq.(4), where $\Delta T = T(z) - T_0$ is temperature rise from the stress-free reference temperature (T_0) which is assumed to exist at a temperature of T_0=0 and $T(z)$ is presented in Eqs. (10)-(12).

The moments and membrane forces include both mechanical and electric components as

$$\begin{aligned} N_r &= N_r^m - N_r^e - N_r^t, \\ N_\theta &= N_\theta^m - N_\theta^e - N_\theta^t, \\ M_r &= M_r^m - M_r^e - M_r^t, \\ M_\theta &= M_\theta^m - M_\theta^e - M_\theta^t \end{aligned} \tag{20}$$

where the superscripts m, e, and t, respectively, denote the mechanical, electric, and temperature components. Mechanical forces and moments of the thin circular plate made of functionally graded material can be expressed as

$$(N_r^m, N_\theta^m) = \int_{-h_f/2}^{h_f/2} (\sigma_{rr}, \sigma_{\theta\theta}) dz \tag{21}$$

$$(M_r^m, M_\theta^m) = \int_{-h_f/2}^{h_f/2} (\sigma_{rr}, \sigma_{\theta\theta}) zdz \tag{22}$$

$$(N_{r\theta}^m, M_{r\theta}^m) = \int_{-h_f/2}^{h_f/2} (1, z) \sigma_{r\theta} dz \tag{23}$$

Substituting Eqs. (13) and (18),(19) into Eqs. (22)-(23) gives the following constitutive relations for mechanical forces and moments of FG plate :

$$\begin{aligned} N_r^m &= D_1(\bar{\varepsilon}_{rr} + \nu \bar{\varepsilon}_{\theta\theta}), \\ N_\theta^m &= D_1(\bar{\varepsilon}_{\theta\theta} + \nu \bar{\varepsilon}_{rr}) \end{aligned} \tag{24}$$

$$\begin{aligned} M_r^m &= D_2(\kappa_{rr} + \nu \kappa_{\theta\theta}), \\ M_\theta^m &= D_2(\kappa_{\theta\theta} + \nu \kappa_{rr}) \end{aligned} \tag{25}$$

$$N_r^t = N_\theta^t = \int_{-h_t/2}^{h_t/2} \frac{\alpha(z) E(z)}{1-\nu} \Delta T(z)\, dz \tag{26}$$

$$M_r^t = M_\theta^t = \int_{-h_t/2}^{h_t/2} \frac{\alpha(z) E(z)}{1-\nu} \Delta T(z)\, zdz \tag{27}$$

in which the coefficients of D_1 and D_2 in the above equations are related to the plate stiffness and are given by

$$D_1 = \int_{-h_f/2}^{h_f/2} \frac{E_f(z)}{1-\nu_f^2} dz, \qquad D_2 = \int_{-h_f/2}^{h_f/2} z^2 \frac{E_f(z)}{1-\nu_f^2} dz \tag{28}$$

It is assumed that the piezoelectric layers are sensitive in both radial and circumferential directions and the piezoelectric permeability constants e_{31}=e_{32}. Hence, the electric membrane forces and bending moments are induced by the converse piezoelectric effect on the piezoelectric actuators, and these forces vary linearly across the plate thickness as [27];

$$N_r^e = N_\theta^e = -e_{31}\left(V_z^t + V_z^b\right)/2, \tag{29}$$

$$M_r^e = M_\theta^e = -e_{31}\left(h_f + h_p\right)\left(V_z^t - V_z^b\right)/2, \tag{30}$$

in which V_z^t and V_z^b are the control voltages applied to the top and bottom piezoelectric layers, respectively.

4.3. System electromechanical equations

Axisymmetric free oscillation equations of the piezoelectric coupled circular FG plate in thermal environment can be derived from the generic piezoelectric shell equations using four system parameters: two Lame parameters, A_1=1, A_2=r, where r is the radial distance measured from the center, and two radii, R_1=∞, R_2=∞ [28,29] as

$$\frac{\partial(rN_r)}{\partial r} - N_\theta = 0 \tag{31}$$

$$\frac{1}{r}\frac{\partial(rQ_{rz})}{\partial r} + N_r\frac{\partial^2 w}{\partial r^2} + N_{\theta\theta}\left(\frac{1}{r}\frac{\partial w}{\partial \theta}\right) - I_1\frac{\partial^2 w}{\partial t^2} = 0 \tag{32}$$

in which $I_1 = \left(\int_{-h_f/2}^{h_f/2} \rho_f(z)dz\right)$ and the transverse shear component Q_{rz} is related to moments as

$$Q_{rz} = \frac{1}{r}\left[\frac{\partial(rM_r)}{\partial r} - M_\theta\right] \tag{33}$$

Note that only the normal radial strain keeps the quadratic nonlinear term. Substituting all force/moment components and strain-displacement equations into the radial and transverse equations (31), (32) yields

$$r\frac{\partial}{\partial r}\left[\frac{1}{r}\frac{\partial}{\partial r}\left(r^2 N_r^m\right)\right] = -\frac{Y}{2}\left(\frac{\partial w}{\partial r}\right)^2 + \frac{\partial}{\partial r}\left[r^2\frac{\partial}{\partial r}\left(N_r^e + N_r^t\right)\right] + \nu r\frac{\partial}{\partial r}\left(N_r^e + N_r^t\right) \tag{34}$$

$$\begin{aligned}&\frac{D_2}{r}\frac{\partial}{\partial r}\left(r\frac{\partial}{\partial r}\left[\frac{1}{r}\frac{\partial}{\partial r}\left(r\frac{\partial}{\partial r}w(r,t)\right)\right]\right) = \\ &-I_1\frac{\partial^2 w}{\partial t^2} + \frac{1}{r}\frac{\partial}{\partial r}\left[r\frac{\partial w}{\partial r}\left(N_r^m - N_r^e - N_r^t\right)\right] - \frac{1}{r}\frac{\partial}{\partial r}\left[r\frac{\partial}{\partial r}\left(M_r^e + M_r^t\right)\right]\end{aligned} \tag{35}$$

in which $Y = \left(\int_{-h_f/2}^{h_f/2} E_f(z)dz\right)$ and boundary conditions at the center of the plate with axisymmetric oscillations are defined as

$$(1)\ \text{Plate center}\ (r=0): Slope: \frac{\partial w}{\partial r} = 0 \tag{36a}$$

$$Radial\ force: N_{rr}^m : finite \tag{36b}$$

Boundary conditions for the simply supported (immovable) circumference are defined as:

Plate circumference ($r=a$):

$$w = 0 \tag{37a}$$

$$\frac{\partial}{\partial r}(rN_r^m) - \nu N_r^m = r\frac{\partial}{\partial r}(N_r^e + N_r^t) \tag{37b}$$

$$-D_2\left(\frac{\partial^2 w}{\partial r^2} + \frac{\nu}{r}\frac{\partial w}{\partial r}\right) = (M_r^e + M_r^t) \tag{37c}$$

It is further assumed that the control potentials on the top and bottom piezoelectric actuators are of equal magnitudes and opposite signs, i.e., $V_z^t = -V_z^b = \hat{V}$ and the plate is subjected to a uniform temperature excitation of *T(z)*. Accordingly, the electric and temperature induced forces and moments can be defined as:

$$N_r^e = N_\theta^e = 0 \tag{38a}$$

$$M_r^e = M_\theta^e = M^e = -e_{31}\left(h_f + h_p\right)\hat{V}, \tag{38b}$$

$$N_\theta^t = N_r^t = N^t, \tag{39a}$$

$$M_\theta^t = M_r^t = M^t, \tag{39b}$$

Using these force and moment expressions, one can further simplify the open-loop plate equations and boundary conditions:

$$r\frac{\partial}{\partial r}\left[\frac{1}{r}\frac{\partial}{\partial r}\left(r^2 N_r^m\right)\right] = -\frac{Y}{2}\left(\frac{\partial w}{\partial r}\right)^2 \tag{40}$$

$$\frac{D_2}{r}\frac{\partial}{\partial r}\left(r\frac{\partial}{\partial r}\left[\frac{1}{r}\frac{\partial}{\partial r}\left(r\frac{\partial}{\partial r}w(r,t)\right)\right]\right) = -I_1\frac{\partial^2 w}{\partial t^2} + \frac{1}{r}\frac{\partial}{\partial r}\left[r\frac{\partial w}{\partial r}\left(N_r^m - N^t\right)\right] \tag{41}$$

Boundary conditions become

Plate center (r=0):

$$Slope: \left.\frac{\partial w}{\partial r}\right|_{r=0} = 0 \tag{42a}$$

$$Radial\ force: \left.N_r^m\right|_{r=0} : finite \tag{42b}$$

Plate circumference (r=a):

$$\left.w\right|_{r=a} = 0 \tag{43a}$$

$$\left[\frac{\partial}{\partial r}(rN_r^m) - \nu N_r^m\right]_{r=a} = 0 \tag{43b}$$

$$\left[-D_2\left(\frac{\partial^2 w}{\partial r^2}+\frac{\nu}{r}\frac{\partial w}{\partial r}\right)\right]_{r=a}=(M^e+M^t) \qquad (43c)$$

4.4. Simplification and Normalization

Solutions of the transverse displacement w and radial force N_r^m of the above open-loop plate equations and boundary conditions can be expressed as a summation of a static component and a dynamic component as

$$w(r,t)=w_s(r)+w_d(r,t) \qquad (44a)$$

$$N_r^m(r,t)=N_{r_s}^m(r,t)+N_{r_d}^m(r,t) \qquad (44b)$$

where $w_s(r)$and $N_{r_s}^m(r,t)$ are the static solutions, $w_d(r,t)$and $N_{r_d}^m(r,t)$ are the dynamic solutions, and the subscripts s and d, respectively, denote the static and dynamic solutions. Accordingly, the solution procedure can be divided into two parts. The first part deals with the nonlinear static solutions, and the second part deals with the dynamic solutions. In addition, normalized dimensionless quantities are adopted in the static and dynamic analyses. These dimensionless quantities are defined by known geometrical and material parameters [30]:

- radial distance:$y=(r/a)^2$,
- transverse deflection:$\bar{w}_s=\sqrt{3(1-\nu^2)}\dfrac{w_s}{h_f}$,
- slope: $X_s(y)=y\dfrac{d\bar{w}_s}{dy}$
- static force: $Y_s^m(y)=\left(a^2N_{r_s}^m/4D_2\right)y$
- radial distance:$x=(r/a)$,
- dynamic deflection:$\bar{w}_d=\sqrt{3(1-\nu^2)}\dfrac{w_d}{h_f}$,
- dynamic force: $Y_d^m(y)=\left(a^2N_{r_d}^m/D_2\right)$
- voltage: $V=[3(1-\nu^2)]^{1/2}e_{31}(h_f+h_p)\hat{V}/(2D_2h_f)$
- temperature load: $T^*=\left(a^2N^t/4D_2\right)$

Substituting these normalized dimensionless quantities into the open-loop plate equations and boundary conditions of axisymmetric plate oscillations and separating the static parts from the dynamic parts gives the static equations and dynamic equations with their associated boundary conditions:

(1) Static Equations *and Boundary Conditions*

$$y^2 \frac{d^2 X_s}{dy^2} = X_s Y_s^m - T^* y X_s \tag{45}$$

$$y^2 \frac{d^2 Y_s^m}{dy^2} = -\frac{1}{2}\left(X_s\right)^2, \qquad 0 < y < 1 \tag{46}$$

Boundary conditions at center y = 0:

$$X_s\big|_{y=0} = 0 \tag{47a}$$

$$Y_s^m\big|_{y=0} = 0 \tag{47b}$$

Boundary conditions on circumference y = 1:

$$\left[\left(1+\nu\right)Y_s^m - 2\frac{dY_s^m}{dy}\right]_{y=1} = 0 \tag{48a}$$

$$\left[\left(1-\nu\right)\frac{X_s}{y} - 2\frac{dX_s^m}{dy}\right]_{y=1} = V\big|_{y=1} \tag{48b}$$

(2) Dynamic Equations and Boundary Conditions

$$x\frac{\partial}{\partial x}\left(\frac{1}{x}\frac{\partial}{\partial x}\left[x^2 Y_d^m\right]\right) = -2\left[\frac{d\overline{w}_s}{dx}\frac{\partial \overline{w}_d}{\partial x} - \frac{1}{2}\left(\frac{\partial \overline{w}_d}{\partial x}\right)^2\right] \tag{49}$$

$$\frac{1}{x}\frac{\partial}{\partial x}\left\{x\frac{\partial}{\partial x}\left[\frac{1}{x}\frac{\partial}{\partial x}\left(x\frac{\partial}{\partial x}(\overline{w}_d)\right)\right]\right\}=$$
$$-\frac{I_1 a^4}{D_2}\frac{\partial^2 \overline{w}_d}{\partial t^2}+\frac{1}{x}\frac{\partial}{\partial x}\left[xY_d^m\frac{d\overline{w}_s}{dx}+\frac{4}{x}Y_s^m\frac{\partial \overline{w}_d}{\partial x}+xY_d^m\frac{\partial \overline{w}_d}{\partial x}\right]-4\hat{T}\frac{1}{x}\frac{\partial}{\partial x}\left(x\frac{\partial \overline{w}_d}{\partial x}\right) \tag{50}$$
$$0<x<1$$

Boundary conditions at center $x = 0$:

$$\left.\frac{\partial \overline{w}_d}{\partial x}\right|_{x=0}=0 \tag{51a}$$

$$\left.Y_d^m\right|_{x=0}=\mathit{finite} \tag{51b}$$

Boundary conditions on circumference $x = 1$:

$$\left.\overline{w}_d\right|_{x=1}=0 \tag{52a}$$

$$\left[\nu\frac{\partial}{\partial x}(Y_d^m)+(1-\nu)Y_d^m\right]_{x=1}=0 \tag{52b}$$

$$\left[\frac{\partial^2 \overline{w}_d}{\partial x^2}+\nu\frac{\partial \overline{w}_d}{\partial x}\right]_{x=1}=0 \tag{52c}$$

5. Static Solutions

For the nonlinear static equations and boundary conditions of the boundary value problem derived above, static solutions of slopes $X_s(y)$and forces $Y_s^m(y)$can be represented in (exact) series expansion forms [30]:

$$X_s(y)=\sum_{i=1}^{\infty}A_i y^i \tag{53a}$$

$$Y_s^m(y) = \sum_{i=1}^{\infty} B_i y^i \qquad 0 \le y \le 1 \tag{53b}$$

where A_i and B_i are constant coefficients. Substituting the series solutions, Eqs. (53a&*b),* into static equations, Eqs. (45) & (46), and grouping coefficients of y^i one can obtain the recurrence equations for coefficients A_i and B_i:

$$A_i = \frac{1}{i(i-1)} \sum_{j=1}^{i-1} A_j B_{i-j} - \hat{T} A_{i-1} \tag{54a}$$

$$B_i = \frac{-1}{2i(i-1)} \sum_{j=1}^{i-1} A_j A_{i-j}, \qquad i = 2, 3, 4, \dots \tag{54b}$$

It is observed that only A_1 and B_1 are independent constants, and the others are dependent. As long as A_1 and B_1 are determined by the boundary conditions, other coefficients A_i and B_i can be calculated from the recurrence equations. Accordingly, static series solutions are completed. The series solutions of $Y_s^m(y)$and $X_s(y)$satisfy the boundary conditions at y = 0, Eqs. (47a,b). Substituting the assumed series solutions$Y_s^m(y)$and $X_s(y)$ into the boundary conditions at y = 1, Eqs. (48a,b), yields

$$\sum_{i=1}^{\infty} \left[(1+\nu-2i) B_i \right] = 0 \tag{55a}$$

$$\sum_{i=1}^{\infty} \left[(1-\nu-2i) A_i \right] = V \tag{55b}$$

A_i and B_i can be determined from the nonlinear algebraic equations Eqs. (55a,b) using the Newton-Raphson iteration method [31]. Define

$$\alpha(A_1, B_1) = \sum_{i=1}^{\infty} \left[(1-\nu-2i) A_i \right] - V \tag{56a}$$

$$\beta(A_1, B_1) = \sum_{i=1}^{\infty} \left[(1+\nu-2i) B_i \right] \tag{56b}$$

$$\overline{A}_1 = A_1 + \Delta_1 \tag{57a}$$

$$\overline{B}_1 = B_1 + \Delta_2 \tag{57b}$$

in which

$$\Delta_1 = \frac{1}{\Delta}\left[\beta(A_1,B_1)\frac{\partial}{\partial B_1}\alpha(A_1,B_1) - \alpha(A_1,B_1)\frac{\partial}{\partial B_1}\beta(A_1,B_1)\right] \tag{58a}$$

$$\Delta_2 = \frac{1}{\Delta}\left[\alpha(A_1,B_1)\frac{\partial}{\partial A_1}\beta(A_1,B_1) - \beta(A_1,B_1)\frac{\partial}{\partial A_1}\alpha(A_1,B_1)\right] \tag{58b}$$

and

$$\Delta = \det\begin{bmatrix} \frac{\partial}{\partial A_1}\alpha(A_1,B_1) & \frac{\partial}{\partial B_1}\alpha(A_1,B_1) \\ \frac{\partial}{\partial A_1}\beta(A_1,B_1) & \frac{\partial}{\partial B_1}\beta(A_1,B_1) \end{bmatrix} \neq 0 \tag{58c}$$

$\overline{A}_1$and $\overline{B}_1$are, respectively, the iteration values of A_1 and B_1 ; $\Delta_1$and Δ_2 are the correction factors of A_1 and B_1 at each iteration. The partial derivatives$\partial\alpha/\partial A_1$,$\partial\alpha/\partial B_1$,$\partial\beta/\partial A_1$ and $\partial\beta/\partial B_1$, can be determined from the definitions of $\alpha(A_1, B_1)$and$\beta(A_1, B_1)$. These iterations are repeated until they reach their prescribed limits, say $|\alpha|$, $|\beta|$, $|\Delta_1|$ and $|\Delta_2|$ are smaller than 10^{-4}. Accordingly, a set of A_1 and B_1 are determined for a set of given control voltages V and temperatures$T$. Using the recurrence equations, one can determine all other A_i's and B_i's, and further the nonlinear static solutions of slope $X_s(y)$ and static force$Y_s^m(y)$. Knowing the slope, one can determine the static deflections $\overline{w}_s$and w_s of the nonlinear circular plate subject to voltage and temperature excitations.

6. Dynamic Solutions

It is assumed that the FG circular plate is oscillating in the vicinity of the nonlinearly deformed static equilibrium position. FG index, voltage and temperature effects to the natural frequencies and amplitude/frequency relations are investigated in this section. First, linearized eigenvalue equations are solved using the exact series solutions. Then nonlinear amplitude and frequency relations of nonlinear large amplitude free vibrations are investigated using the Galerkin method and the perturbation method.

6.1. Eigenvalue Equations

Neglect the nonlinear terms in the normalized dynamic equations, and then assume following harmonic solutions of displacement and dynamic force

$$\bar{w}_d(x,t) = R_d(x)\sin(\omega_n t) \tag{59a}$$

$$Y_d^m(x,t) = S_d(x)\sin(\omega_n t) \tag{59b}$$

where ω_n is the natural frequency; $R_d(x)$and $S_d(x)$are the (linear) eigenfunctions or mode shape functions of $\bar{w}_d(x, t)$and$Y_d^m(x, t)$, respectively. $R_d(x)$defines the mode shape function, and $S_d(x)$ defines the spatial force distribution. Both $R_d(x)$and $S_d(x)$ have to satisfy the boundary conditions, and they are also assumed in the series expansion forms. Substituting Eqs. (59a, b) into the dynamic equations and boundary conditions, Eqs. (49)-(52), yields

$$x\frac{d}{dx}\left(\frac{1}{x}\frac{d}{dx}\left[x^2 S_d(x)\right]\right) = -2\frac{d\bar{w}_s}{dx}\frac{dR_d}{dx} \tag{60}$$

$$\begin{aligned}&\frac{1}{x}\frac{d}{dx}\left\{x\frac{d}{dx}\left[\frac{1}{x}\frac{d}{dx}\left(x\frac{d}{dx}(R_d(x))\right)\right]\right\} = \\ &\lambda R_d(x) + \frac{1}{x}\frac{d}{dx}\left[2xS_d(x)\frac{d\bar{w}_s}{dx} + \frac{4}{x}Y_s^m\frac{dR_d}{dx}\right] - 4\hat{T}\frac{1}{x}\frac{d}{dx}\left[x\frac{dR_d}{dx}\right],\ 0<x<1\end{aligned} \tag{61}$$

where λ is the eigenvalue and$\lambda = I_1\frac{a^4}{D_2}\omega_n^2$. Boundary conditions become

1. Center x=0:

$$\left.\frac{dR_d}{dx}\right|_{x=0} = 0 \tag{62a}$$

$$\left.S_d(x)\right|_{x=0} : finite \tag{62b}$$

2. Circumference x=1:

$$R_d\big|_{x=1} = 0 \tag{63a}$$

$$\left[\nu \frac{d}{dx}[S_d(x)] + (1-\nu) S_d(x)\right]_{x=1} = 0 \tag{63b}$$

$$\left[\frac{d^2 R_d}{dx^2} + \nu \frac{dR_d}{dx}\right]_{x=1} = 0 \tag{63c}$$

Again, assume the eigenfunctions take the series expansion forms:

$$R_d(x) = \sum_{i=0}^{\infty} a_i x^{2i} \tag{64a}$$

$$S_d(x) = \sum_{i=0}^{\infty} b_i x^{2i} \tag{64b}$$

where a_i andb_i, are constants determined by eigenvalue equations and boundary conditions. The series solutions $R_d(x)$ and$S_d(x)$satisfy the boundary conditions at x = 0, Eqs. (62a,b). Assume a_i andb_i, be represented by the linear combinations of independent constantsa_0, $a_1$and b_0.

$$a_i = f_{i1} a_0 + f_{i2} a_1 + f_{i3} b_0 \tag{65a}$$

$$b_i = g_{i1} a_0 + g_{i2} a_1 + g_{i3} b_0\,, \quad i = 1, 2, 3, ... \tag{65b}$$

where f_{ij} and $g_{ij}$are to be determined. Substituting the series expressions of modes $R_d(x)$and forces $S_d(x)$ into the dynamic equations, one can derive a set of recurrence equations of a_i andb_i. Then, using expressions of a_i and b_i of Eqs. (65a, b), one can further determine the coefficients f_{ij} and g_{ij} $f_{01}=1$, $f_{02}=f_{03}=0$; $f_{11}=0$, $f_{12}=1$, $f_{13}=0$; $g_{ik} = -\frac{2}{i(i+1)}\sum_{j=1}^{i} j f_{jk} A_{i-j+1}$, $i=1, 2, 3, ...$, $k=1, 2, 3.$,$g_{01}=0$, $g_{02}=0$, $g_{03}=1$; $f_{21}=\lambda/64$, $f_{23}=A_1/8$, $f_{33}=[A_1 g_{13} + 4 f_{23}(B_1 - T^*) + A_2]/36$, $f_{32}=\{\lambda + 32B_2 + 16[A_1 g_{12} + 4 f_{22}(B_1 - T^*)]\}/24$, $f_{31}=[A_1 g_{11} + 4 f_{21}(B_1 - T^*)]/36$,$f_{22}=(B_1 - T^*)/4$

$f_{(i+2)k} = \left[\lambda f_{ik} - 16T^*(i+1)^2 + 8(i+1) \times \sum_{j=1}^{i+1} (2 j f_{ik} B_{i-k+2} + A_j g_{(i-j+1)k})\right][4(i+1)(i+2)]^{-2}$ $i = 2, 3, 4, \ldots,$ and $k = 1, 2, 3$.

Substituting these coefficients into the boundary conditions at x = 1, one can obtain an explicit matrix representation of the eigenvalue equation.

$$\begin{bmatrix} h_{11} & h_{12} & h_{13} \\ h_{21} & h_{22} & h_{23} \\ h_{31} & h_{32} & h_{33} \end{bmatrix} \begin{bmatrix} a_0 \\ a_1 \\ b_0 \end{bmatrix} = [0] \tag{66}$$

where h_{ij} are defined by

$$h_{1k} = \sum_{i=0i1}^{\infty} f_{ik} \tag{67a}$$

$$h_{2k} = \sum_{i=0}^{\infty} \left[i\nu + i(2i-1) \right] f_{ik} \quad , \quad k = 1,2,3. \tag{67b}$$

$$h_{3k} = \sum_{i=0}^{\infty} \left[2i\nu + (1-\nu) \right] g_{ik} \quad , \quad k = 1,2,3. \tag{67c}$$

These h_{ik} coefficients are functions of eigenvalues λ, and accordingly, the determinant of the coefficient matrix leads to a nonlinear characteristic equation. Using the Newton-Raphson iteration method [31], one can calculate eigenvalues and furthermore natural frequencies and mode shape functions of the nonlinear FG circular plate.

6.2. Nonlinear Large Amplitude Free Vibrations

In this section, the perturbation method is used to investigate the nonlinear large amplitude effect to natural frequencies of the piezoelectric laminated FG circular plate. Assume an approximate solution of the nonlinear response $\bar{w}_d(x, t)$ be a product of a spatial function $\bar{w}_d^*(x)$ and a temporal function $f(t)$;

$$\bar{w}_d(x,t) = \bar{w}_d^*(x) f(t) \tag{68}$$

where $\overline{w}_d^*(x)$ is a test function satisfying the boundary conditions:

1. Plate center x=0: $\left.\frac{\partial \overline{w}_d^*}{\partial x}\right|_{x=0}=0$

2. Plate circumference x=1: $\overline{w}_d^*\Big|_{x=1}=0$,

$$\left[\frac{\partial^2 \overline{w}_d^*}{\partial x^2}+\nu\frac{\partial \overline{w}_d^*}{\partial x}\right]_{x=1}=0 \tag{69}$$

Substituting Eq. (68) into the dynamic equation Eq. (49) and using the Galerkin method, one can derive a nonlinear equation for $f(\tau)$

$$f_{,\tau\tau}+f(\tau)+\mu_1 f^2(\tau)+\mu_2 f^3(\tau)=0 \tag{70}$$

where $\tau=\omega t$ and μ_1 and μ_2 are the nonlinear coefficient functions, ω is defined by

$$\omega^2=(c_1+c_3)/c_2. \tag{71}$$

c_1,c_2 and c_3 are defined by integrals:

$$c_1=\int_0^1\left\{\overline{w}_d^*(x)\frac{d}{dx}\left\{x\frac{d}{dx}\left[\frac{1}{x}\frac{d}{dx}\left(x\frac{d}{dx}(\overline{w}_d^*(x))\right)\right]\right\}\right\}dx \tag{72a}$$

$$c_2=\int_0^2\left(\overline{w}_d^*(x)\right)^2 x dx \tag{72b}$$

$$c_3=-\int_0^1\overline{w}_d^*(x)\frac{d}{dx}\left[\frac{4}{x}Y_s^m\frac{d\overline{w}_d^*}{dx}+2xN_{r_d}^{*l}\frac{d\overline{w}_s}{dx}-4T*x\frac{d\overline{w}_d^*}{dx}\right]dx \tag{72c}$$

in which $N_{r_d}^{*l}(x)$ and $N_{r_d}^{*n}(x)$ are the linear and nonlinear force components. The nonlinear coefficient functions μ_1 and μ_1 are defined by

$$\mu_1=c_4/(c_1+c_3) \tag{73a}$$

$$\mu_2 = c_5/(c_1 + c_3) \tag{73b}$$

where c_4 and $c_5$are defined by integrals:

$$c_4 = -\int_0^1 \overline{w}_d^*(x)\frac{d}{dx}\left[2xN_{r_d}^{*n}(x)\frac{d\overline{w}_s(x)}{dx} + N_{r_d}^{*l}(x)\frac{d\overline{w}_d^*(x)}{dx}\right]dx \tag{74a}$$

$$c_5 = -\int_0^1 \overline{w}_d^*(x)\frac{d}{dx}\left[xN_{r_d}^{*n}(x)\frac{d\overline{w}_d^*(x)}{dx}\right]dx \tag{74b}$$

The linear and nonlinear force components are written as

$$N_{r_d}^{*l}(x) = \frac{1}{x^2}\int_0^1 G(x,\xi)\frac{1}{\xi}\frac{d\overline{w}_s}{d\xi}\frac{d\overline{w}_d^*}{d\xi}d\xi \tag{75a}$$

$$\overline{N}_{r_d}^{*n}(x) = \frac{1}{4x^2}\int_0^1 G(x,\xi)\frac{1}{\xi}\left(\frac{d\overline{w}_d^*}{d\xi}\right)^2 d\xi \tag{75b}$$

The kernel function

$$G(x,\xi) = \begin{cases} \left[1-((1+\nu)/(1-\nu))\xi^2\right]x^2 & x < \xi, \\ \left[1-((1+\nu)/(1-\nu))x^2\right]\xi^2 & x \geq \xi, \end{cases} \tag{76}$$

Since the mode shape functions of the linear free vibrations were determined previously, it is convenient to select the mode shape functions as the trial functions,

$$\overline{w}_d^*(x) = R_d(x), \tag{77a}$$

$$N_{r_d}^{*l}(x) = S_d(x), \quad 0 \leq x \leq 1; \tag{77b}$$

and the frequency can be selected as the natural frequencies determined previously. Using the Krylov-Bogoliubov-Mitropolsky perturbation method [32] and solving the nonlinear dynamic equation Eq. (70), one can obtain the amplitude-frequency relation as

$$p = \frac{\omega}{\omega_n} = 1 + \frac{1}{24}\left(9\mu_2 - 10\mu_1^2\right)\hat{a}^2 + \frac{5}{8}\mu_1\mu_2\hat{a}^3 + \ldots \quad (78)$$

where ω is the (nonlinear vibration) frequency; ω_n is the natural frequency; $\hat{a}$ is the dimensionless vibration amplitude; and μ_1 and μ_2 are the nonlinear coefficient functions. Note that the ratio is unity, i.e., $p=\omega / \omega_n=1$, if the system is linear. Once the coefficients μ_1 and μ_2 are calculated, one can further evaluate the amplitude and frequency relations of the nonlinear circular plate subjected to temperature excitations and control voltages.

7. Results and Discussions

Temperature effects of nonlinear static deformations, control voltages, and linear and nonlinear free vibrations of a simply supported piezoelectric laminated functionally graded circular plate are investigated in this section.

7.1. Comparison studies

To ensure the accuracy of the present analysis, an illustrative example is solved. The relevant material properties are listed in Table 1. Since there are no appropriate comparison results available for the problems being analyzed in this chapter, we decided to verify the validity of the obtained results by comparing with those of the FEM results. Our FEM model for piezo-FG plate consists of a 3D 8-noded solid element with number of total nodes 26950, number of total elements 24276, 4 DOF per node (3 translation, temperature) in the host plate element and 6 DOF per node (3 translation, temperature, voltage and magnetic properties) in the piezoelectric element. The finite element model has been programmed by the authors, while the standard bilinear interpolations have been employed in finite element approximations.

Table 2 compares the present results of normalized dimensionless central deflections $W_s=\sqrt{3(1-\nu^2)}(w_s / h_f)$ with finite element solutions in analyzing the effect of normalized dimensionless piezoelectric voltages $V=[3(1-\nu^2)]^{1/2} e_{31}(h_f+h_p)V^* / (2D_2 h_f)$ to the normalized dimensionless center deflections at various normalized temperatures $(T^*=(a^2 N^t / 4D_2))$ in which a nonlinear deflection-voltage relationship can be observed. As seen from Table 2 the maximum estimated difference of the proposed solution with finite element method is about 0.079%, and a close correlation between these results validates the proposed method of solution.

In general, a higher temperature induces higher deflections of the plate, and the deflection at each temperature is attenuated when the control voltage increases and the effect of imposed voltage on the center deflection is nonlinear and this effect is predominant in lesser voltage amounts. This effect can also be seen in the case of considering the temperature environment effect. For example, when $T^* = 0.2$by increasing the imposed voltage from 0.6 to 1.2 (100%) the normalized dimensionless center deflections increases about 46.8%, while it increases about 34.5% when the imposed voltage increases from 1.2 to 2.4 (100%). In the case of $T^* = 0.5$by increasing the imposed voltage from 0.6 to 1.2 (100%) the normalized dimensionless center deflection increases about 36.5% while it increases about 28.7% when the imposed voltage increases from 1.2 to 2.4 (100%).

Material	Property					
	E (GPa)	ρ (kg/m^3)	ν	α ($1/°C$)	κ (W/mK)	d_{31}, d_{32} (m/V)
Aluminum	70	2707	0.3	23e-6	204	-
Alumina	380	3800	0.3	7.4e-6	10.4	-
PZT	63	7600	0.3	1.2e-4	0.17	1.79e-10

Table 1. Material properties [13].

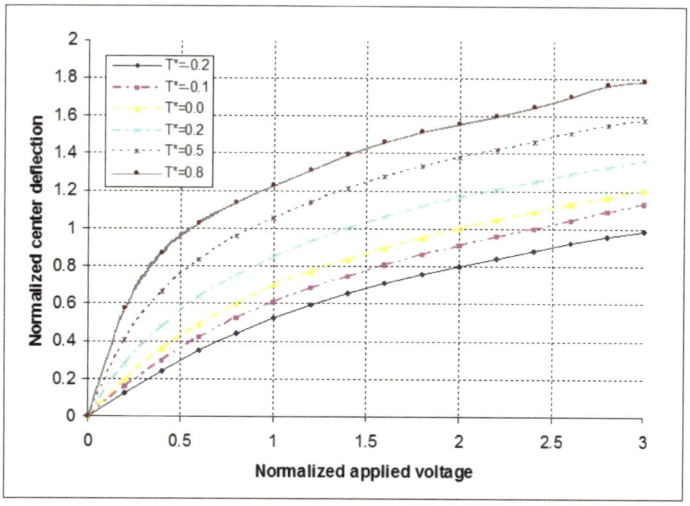

Figure 2. Effect of applied voltage to the normalized center deflection for various normalized temperatures $T^* = (a^2 N^t / 4D_2)$ (Metal plate)

Normalized Voltage (V)	Normalized Temperature(T^*)					
	$T^*=0$			$T^*=0.2$		
	Present	FEM	Diff. (%)	Present	FEM	Diff. (%)
0	0	0	0	0	0	0
0.4	0.3537	0.3538	0.041	0.8659	0.8663	0.044
0.8	0.5982	0.5985	0.048	1.1347	1.1352	0.046
1.2	0.7681	0.7685	0.055	1.3118	1.3126	0.060
1.6	0.8925	0.8930	0.057	1.4601	1.4610	0.062
2	0.9924	0.9930	0.065	1.5581	1.5592	0.070
2.4	1.0777	1.0784	0.068	1.6458	1.6470	0.074
2.8	1.1509	1.1517	0.071	1.7653	1.7666	0.076
	$T^*=0.5$			$T^*=0.8$		
	Present	FEM	Diff. (%)	Present	FEM	Diff. (%)
0	0	0	0	0	0	0
0.4	0.6633	0.6636	0.043	0.8659	0.8663	0.044
0.8	0.9569	0.9573	0.045	1.1347	1.1352	0.046
1.2	1.1380	1.1387	0.058	1.3118	1.3126	0.060
1.6	1.2746	1.2754	0.060	1.4601	1.4610	0.062
2.0	1.3776	1.3785	0.068	1.5581	1.5592	0.070
2.4	1.4643	1.4653	0.072	1.6458	1.6470	0.074
2.8	1.5510	1.5522	0.075	1.7653	1.7666	0.076

Table 2. Values of the normalized dimensionless center deflections with respect to the normalized dimensionless piezoelectric voltages for various normalized temperatures computed by two methods (present series solution and FEM) (v=0.3, n=1000)

7.2. Parametric studies

Having validated the foregoing formulations, we began to study the large amplitude vibration behavior of FG laminated circular plate subjected to thermo-electro-mechanical loading. The results for laminated plates with isotropic substrate layers (that is, the substrate is purely metallic or purely ceramic) and with graded substrate layers (various n) are given in both tabular and graphical forms.

To investigate the effect of the applied actuator voltage on the non-linear thermo-electromechanical vibration, the nonlinear normalized center deflection of various graded plates under various applied normalized voltages is tabulated in Table 3.

Normalized Temp. (T^*)	FGM index (n) / Normalized Voltage (V)							
	Metal				n=10			
	V=0.2	V=0.3	V=0.5	V=1	V=0.2	V=0.3	V=0.5	V=1
0.0	0.1820	0.2724	0.3924	0.6925	0.1206	0.1805	0.2601	0.4590
0.2	0.2738	0.3869	0.5213	0.8573	0.1815	0.2564	0.3455	0.5681
0.4	0.3654	0.5010	0.6421	0.9949	0.2422	0.3320	0.4255	0.6593
0.6	0.4706	0.6169	0.7596	1.1164	0.3119	0.4089	0.5034	0.7399
0.8	0.6071	0.7400	0.8808	1.2329	0.4024	0.4905	0.5838	0.8171
(T^*)					n=0.5			
	V=0.2	V=0.3	V=0.5	V=1	V=0.2	V=0.3	V=0.5	V=1
0.0	0.0992	0.1484	0.2138	0.3773	0.0939	0.1405	0.2024	0.3572
0.2	0.1492	0.2108	0.2840	0.4670	0.1413	0.1996	0.2689	0.4422
0.4	0.1991	0.2729	0.3498	0.5420	0.1885	0.2584	0.3312	0.5132
0.6	0.2564	0.3361	0.4138	0.6082	0.2428	0.3183	0.3919	0.5759
0.8	0.3308	0.4032	0.4799	0.6716	0.3132	0.3818	0.4544	0.6360
(T^*)					Ceramic (n=0)			
	V=0.2	V=0.3	V=0.5	V=1	V=0.2	V=0.3	V=0.5	V=1
0.0	0.0838	0.1255	0.1807	0.3190	0.0716	0.1072	0.1544	0.2725
0.2	0.1261	0.1782	0.2401	0.3948	0.1078	0.1522	0.2051	0.3373
0.4	0.1683	0.2307	0.2957	0.4582	0.1438	0.1971	0.2527	0.3915
0.6	0.2168	0.2841	0.3499	0.5142	0.1852	0.2428	0.2989	0.4393
0.8	0.2796	0.3409	0.4057	0.5678	0.2389	0.2912	0.3466	0.4851

Table 3. FGM index and normalized voltage effects to the nonlinear center deflection

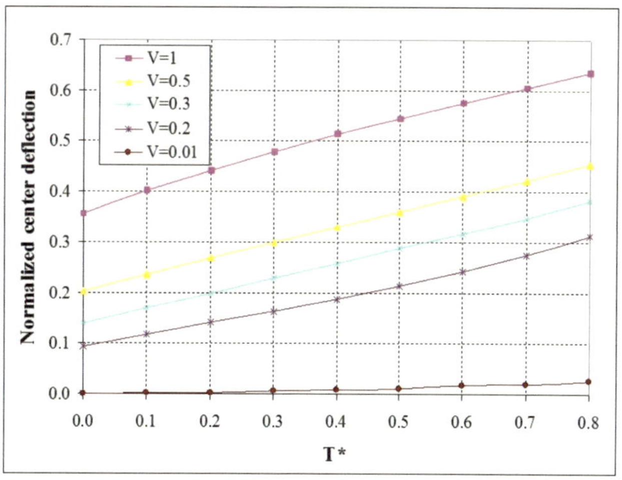

Figure 3. Normalized Temperature effects to the center deflection for various values of Voltages (n=0.5)

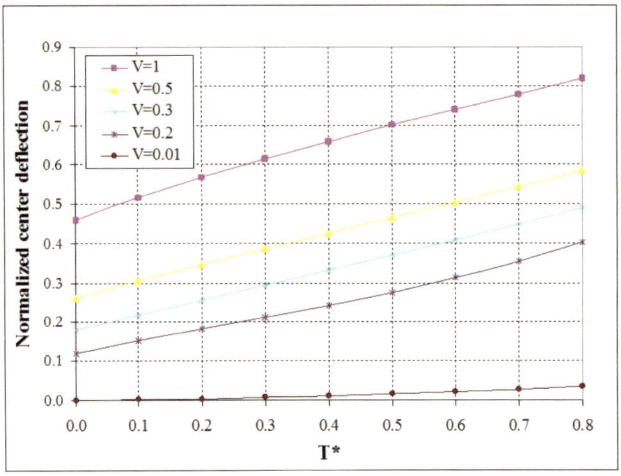

Figure 4. Normalized Temperature effects to the center deflection for various values of Voltages (n=10)

For instance, Figs. 3 and 4. depict the normalized temperature and voltage effects on the center deflection of two graded plates (n=0.5 n=10). It shows that increasing the normalized temperature makes the center deflection increase in various voltages, but this effect is predominant at higher voltages. Figures 5~6 depict the effect of FGM index on the non-linear thermo-electro-mechanical behavior (center deflection) of FGM plates with different normalized applied voltages in logarithmic scale. It is also obvious from these figures that, by increasing the material gradients, the normalized center deflection would be increased in various temperature fields, and it is also demonstrated that larger thermal gradients will lead to greater deflections. This trend can be seen in various material gradients, which means that the non-linear deflection can be controlled by applying the appropriate voltage in the piezoelectric actuator layers.

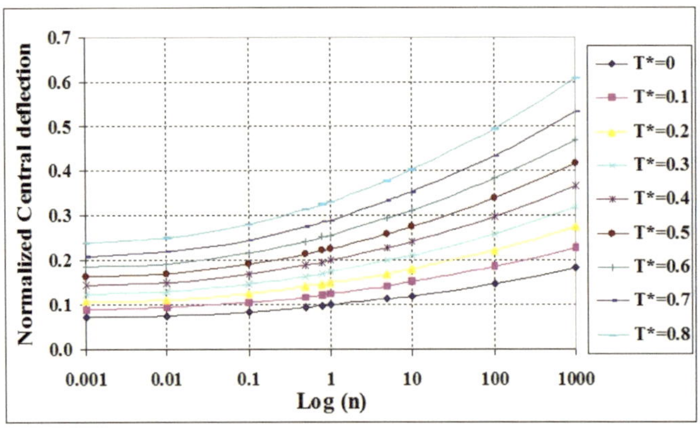

Figure 5. FGM index effects on nonlinear center deflection for various normalized temperature (V=0.2)

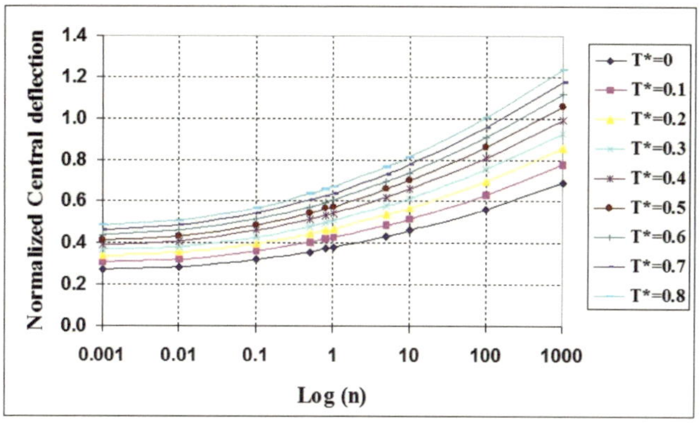

Figure 6. FGM index effects on nonlinear center deflection for various normalized temperature (V=1)

Normalized Voltage (V)	FGM index (n) / Normalized Temp. (T*)							
	Metal				n=10			
	Normalized Temperature (T*)				Normalized Temperature(T*)			
	0	0.2	0.5	0.8	0	0.2	0.5	0.8
0.0	*4.8891*	*4.4249*	*3.5293*	*2.3669*	*5.5359*	*5.0103*	*3.9962*	*2.6800*
0.4	*5.5796*	*5.67452*	*5.92735*	*6.49712*	*6.3178*	*6.4253*	*6.7116*	*7.3567*
0.8	*6.7141*	*7.26308*	*8.00595*	*8.71889*	*7.6024*	*8.2240*	*9.0652*	*9.8724*
1.2	*7.7799*	*8.452*	*9.37002*	*10.1564*	*8.8092*	*9.5702*	*10.6097*	*11.5001*
1.6	*8.6394*	*9.33903*	*10.3511*	*11.2809*	*9.7824*	*10.5746*	*11.7206*	*12.7734*
2.0	*9.3329*	*10.1399*	*11.2449*	*12.2285*	*10.5677*	*11.4814*	*12.7326*	*13.8464*
2.4	*9.9414*	*10.8734*	*12.0351*	*13.0241*	*11.2567*	*12.3120*	*13.6274*	*14.7472*
2.8	*10.5104*	*11.4499*	*12.6036*	*13.7116*	*11.9010*	*12.9647*	*14.2711*	*15.5257*
	0	0.2	0.5	0.8	0	0.2	0.5	0.8
0.0	*7.4121*	*6.7083*	*5.3506*	*3.5883*	*11.2882*	*10.2165*	*8.1487*	*5.4648*
0.4	*8.4589*	*8.6028*	*8.9861*	*9.8499*	*12.8825*	*13.1017*	*13.6854*	*15.0009*
0.8	*10.1789*	*11.0111*	*12.1374*	*13.2182*	*15.5020*	*16.7694*	*18.4846*	*20.1307*
1.2	*11.7947*	*12.8136*	*14.2054*	*15.3975*	*17.9627*	*19.5145*	*21.6340*	*23.4497*
1.6	*13.0977*	*14.1584*	*15.6927*	*17.1023*	*19.9472*	*21.5625*	*23.8992*	*26.0460*
2.0	*14.1491*	*15.3725*	*17.0478*	*18.5389*	*21.5483*	*23.4116*	*25.9629*	*28.2339*
2.4	*15.0716*	*16.4845*	*18.2458*	*19.7451*	*22.9533*	*25.1051*	*27.7874*	*30.0708*
2.8	*15.9342*	*17.3585*	*19.1076*	*20.7874*	*24.2670*	*26.4362*	*29.0999*	*31.6582*

Table 4. FGM index and normalized temperature effects to the first natural frequency for various normalized voltages.

We examine in this section the effect of control voltages and thermal environment on the vibration characteristics of the piezoelectric laminated circular FG plate for various FGM indexes. To this end, Table 4 as well as the Figures 4.1~4.3 show the nonlinear relationships between the first natural frequencies$\omega_i a^2\sqrt{I_1 / D_2}$, versus the normalized temperature in various normalized control voltages$V$. These free vibrations are assumed to be in the vicinity of the nonlinearly deformed static equilibrium position.

Also, the effect of normalized temperature on the first natural frequency of the FG circular plate for various FGM indexes under various normalized control voltage is investigated and tabulated in Table 4, while the voltage-dependent first natural frequency changes are plotted in Figures 7~9 for various temperatures. It is seen that the imposed voltage has a significant effect on the first natural frequency of the structure, and by increasing the imposed voltage, the first natural frequency increases in a nonlinear manner. For instance, for the FGM plate with n=10 by increasing the imposed voltage from 0 to 0.2 the first natural frequency increases about 4.84%, while by increasing the voltage from 0.2 to 0.3 the first natural frequency increases about 15.12%.

It is seen that the imposed thermal environment has a significant effect on the first natural frequency of the structure, and by increasing the imposed temperature, the first natural frequency decreases in a nonlinear manner. However, this thermal tendency of decreasing the natural frequency can be compensated and corrected with the control voltages V, as shown in Fig. 7. ~ 8.

Frequency variations of large amplitude oscillations with temperature and applied voltage changes are also investigated and plotted in Figure 6. There are two sets of curves in this figure. The first set has no control voltages, and the second set has control voltages. It is observed that the control voltages actually reduce the nonlinear frequency and amplitude ratios, i.e.,$P \rightarrow 1$. Accordingly, the nonlinear frequency and amplitude ratios can be actively controlled and the nonlinear effects reduced, i.e., the ratio is approaching to 1- the linear case. Other studies of the second natural frequency also suggest that the second natural frequency exhibits very much similar phenomena of the first natural frequency.

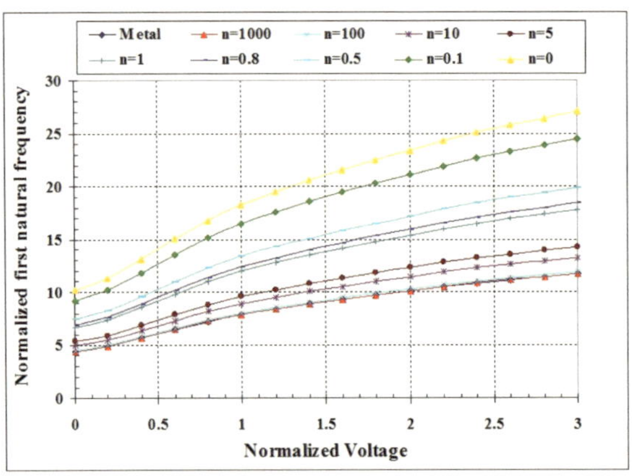

Figure 7. Effect of normalized voltage to first natural frequency for various FGM indexes (T*=0.2)

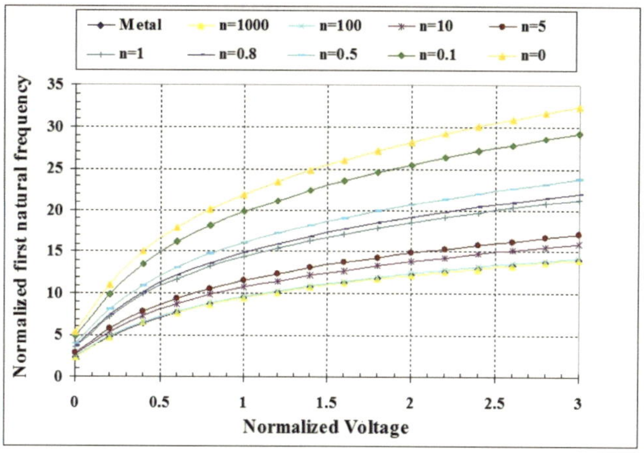

Figure 8. Effect of normalized voltage to first natural frequency for various FGM indexes (T*=0.8)

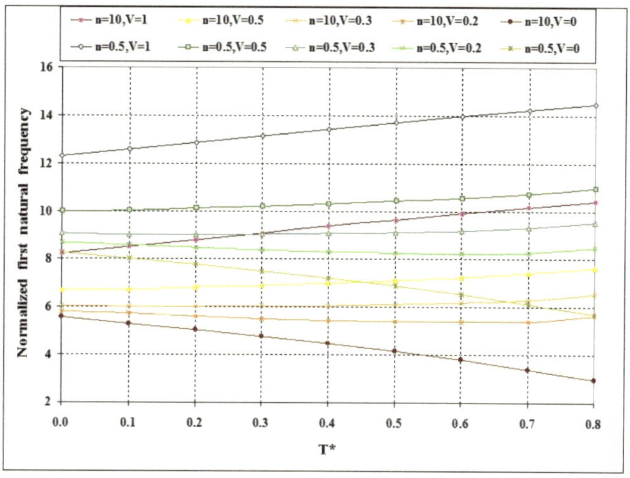

Figure 9. Effects of normalized temperature on the normalized first natural frequency for various normalized voltages (n=10,n=0.5)

8. Summary and Conclusions

A piezoelectric bounded circular FG plate subjected to temperature changes and control voltages is investigated based on classical plate theory, including the effects of the thermal gradient, piezothermoelasticity and von Karman type geometric nonlinearity. Nonlinear coupled open-loop plate equations in radial and transverse oscillations were derived first, and then the equations were simplified to an axisymmetric oscillation case. An exact solution technique based on series-type solutions is used to obtain piezothermoelastic solutions for nonlinear static deformations and natural frequencies of the FG circular plate subjected to temperature and voltage excitations. Voltage controlled natural frequencies of the first mode at various temperatures are studied. It is observed that a higher temperature induces higher deflections of the plate, and the deflection at each temperature is attenuated when the control voltage increases, but this effect is predominant in higher voltages. Also by increasing the FGM gradient index the normalized center deflection will increase in a nonlinear manner in various temperature fields. It is seen that the imposed thermal environment has a significant effect on the natural frequency of the structure, and by increasing the imposed temperature, the natural frequency decreases in a nonlinear manner for various FGM indexes; this effect is predominant at higher temperatures. Both the nonlinear static deflections and natural frequencies are influenced by the temperatures and control voltages geometric and the static control voltages can be used to compensate nonlinear deflections.

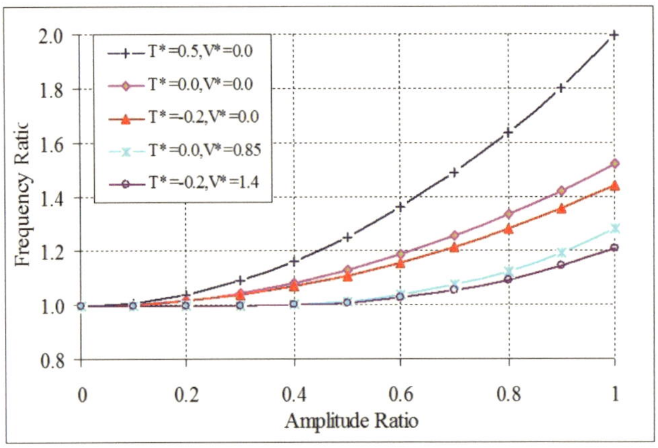

Figure 10. Temperature/control effects on amplitude dependent first natural frequency (for metal plate) – amplitude ratio: w/h_f, frequency ratio: (ω / ω_1)

Appendix A:

$$T_1 = T_2 + (T_U - T_L) - \frac{\kappa_c d + \kappa_m}{\kappa_m}(T_2 - T_L), \quad T_2 = T_L + \left[\frac{\kappa_m}{c h_f}(T_U - T_L)\right] \Big/ \left[\frac{\kappa_p}{h_p} + \frac{\kappa_c d + \kappa_m}{c h_f}\right]$$

$$A_0 = T_2, \qquad A_1 = \frac{T_1 - T_2}{c} \qquad\qquad A_2 = -\frac{T_1 - T_2}{c}\frac{\kappa_{cm}}{(N+1)\kappa_m}$$

$$A_3 = \frac{T_1 - T_2}{c}\frac{\kappa_{cm}^2}{(2N+1)\kappa_m^2} \qquad\qquad A_4 = -\frac{T_1 - T_2}{c}\frac{\kappa_{cm}^3}{(3N+1)\kappa_m^3}$$

$$A_5 = \frac{T_1 - T_2}{c}\frac{\kappa_{cm}^4}{(4N+1)\kappa_m^4} \qquad\qquad A_6 = -\frac{T_1 - T_2}{c}\frac{\kappa_{cm}^5}{(5N+1)\kappa_m^5}$$

where

$$\kappa_{cm}=\kappa_c-\kappa_m$$

$$c=1-\frac{1}{N+1}\frac{\kappa_{cm}}{\kappa_m}+\frac{1}{2N+1}\left(\frac{\kappa_{cm}}{\kappa_m}\right)^2-\frac{1}{3N+1}\left(\frac{\kappa_{cm}}{\kappa_m}\right)^3+\frac{1}{4N+1}\left(\frac{\kappa_{cm}}{\kappa_m}\right)^4-\frac{1}{5N+1}\left(\frac{\kappa_{cm}}{\kappa_m}\right)^5$$

$$d=1-\frac{\kappa_{cm}}{\kappa_m}+\left(\frac{\kappa_{cm}}{\kappa_m}\right)^2-\left(\frac{\kappa_{cm}}{\kappa_m}\right)^3+\left(\frac{\kappa_{cm}}{\kappa_m}\right)^4-\left(\frac{\kappa_{cm}}{\kappa_m}\right)^5$$

Acknowledgements

The work described in this chapter was funded by a grant from International University of Imam Khomeini (Grant No. 385022-1391) The author is grateful for this financial support.

Author details

Farzad Ebrahimi*

Address all correspondence to: f.ebrahimi@ikiu.ac.ir

Department of Mechanical Engineering, Faculty of Engineering, Imam Khomeini International University, Qazvin, Iran

References

[1] Peng, F., Ng, A., & Hu, Y. R. (2005). Actuator placement optimization and adaptive vibration control of plate smart structures. J. Intell. Mater. Syst. Struct. , 16-263.

[2] Dong, S., & Tong, L. (2001). Vibration control of plates using discretely distributed piezoelectric quasi-modal actuators/sensors. *AIAA J.*, 39-1766.

[3] Ray, M. C. (2003). Optimal control of laminated shells with piezoelectric sensor and actuator layers,. *AIAA J.*, 41-1151.

[4] Spencer, W. J., Corbett, W. T., Dominguez, L. R., & Shafer, B. D. (1978). An electronically controlled piezoelectric insulin pump and valves. *IEEE Trans. Sonics Ultrason.*, 25-153.

[5] Dong, S., Du, X., Bouchilloux, P., & Uchino, K. (2002). Piezoelectric ring-morph actuation for valve application. *J. Electroceram*, 8-155.

[6] Chee, C. Y. K., Tong, L., & Steve, G. P. (1998). A review on the modeling of piezoelectric sensors and actuators incorporated in intelligent structures,. *J. Intell. Mater. Syst. Struct.*, 9-3.

[7] Cao, L., Mantell, S., & Polla, D. (2001). Design and simulation of an implantable medical drug delivery system using microelectromechanical systems technology. *Sensors Actuators A*, 94-117.

[8] Chen, X., Fox, C. H. J., & Mc William, S. (2004). Optimization of a cantilever microswitch with piezoelectric actuation. *J. Intell. Mater. Syst. Struct.*, 15-823.

[9] Reddy, J. N., & Cheng, Z. Q. (2001). Three-dimensional solutions of smart functionally graded plates. *ASME J. Appl. Mech.*, 68-234.

[10] Wang, B. L., & Noda, N. (2001). Design of smart functionally graded thermo-piezoelectric composite structure. *Smart Mater. Struct.*, 10-189.

[11] He, X. Q., Ng, T. Y., Sivashanker, S., & Liew, K. M. (2001). Active control of FGM plates with integrated piezoelectric sensors and actuators. , Int. J. Solids Struct. , 38-1641.

[12] Yang, J., Kitipornchai, S., & Liew, K. M. (2004). Nonlinear analysis of thermo-electro-mechanical behavior of shear deformable FGM plates with piezoelectric actuators. *Int. J. Numer. Methods Eng.*, 59-1605.

[13] Huang, X. L., & Shen, H. S. (2006). Vibration and dynamic response of functionally graded plates with piezoelectric actuators in thermal environments. *J. Sound Vib.*, 289-25.

[14] Liew, K. M., He, X. Q., Ng, T. Y., & Kitipornchai, S. (2003). Finite element piezothermoelasticity analysis and the active control of FGM plates with integrated piezoelectric sensors and actuators. *Computational Mechanics*, 31-350.

[15] Zhai, P. C., Zhang, Q. J., & Yuan, R. Z. (1997). Influence of temperature dependent properties on temperature response and optimum design of C/M FGMs under thermal cyclic loading. *J. Wuhan Univ. Technol.*, 12-19.

[16] Ebrahimi, F., & Rastgoo, A. (2008). Free vibration analysis of smart annular FGM plates integrated with piezoelectric layers. *Smart Mater. Struct,*, 17(015044).

[17] Ebrahimi, F., & A.and, Rastgoo. (2008). An analytical study on the free vibration of smart circular thin FGM plate based on classical plate theory. *Thin-Walled Structures*, 46-1402.

[18] Ebrahimi, F., & Rastgoo, A. (2008). Free Vibration Analysis of Smart FGM Plates. *International Journal of Mechanical Systems Science and Engineering*, 2(2), 94-99.

[19] Ebrahimi, F., Rastgoo, A., & Kargarnovin, M. H. (2008). Analytical investigation on axisymmetric free vibrations of moderately thick circular functionally graded plate integrated with piezoelectric layers. *Journal of Mechanical Science and Technology*, 22(6), 1056-1072.

[20] Ebrahimi, F., Rastgoo, A., & Atai, A. A. (2009). Theoretical analysis of smart moderately thick shear deformable annular functionally graded plate. *European Journal of Mechanics- A/Solids*, 28-962.

[21] Ebrahimi, F., & Rastgoo, A. (2009). Nonlinear vibration of smart circular functionally graded plates coupled with piezoelectric layers. *Int J Mech Mater Des*, 5-157.

[22] Reddy, J. N., & Praveen, G. N. (1998). Nonlinear transient thermoelastic analysis of functionally graded ceramic-metal plate,. *Int. J. Solids Struct.*, 35-4457.

[23] Wetherhold, R. C., Seelman, S., & Wang, S. (1996). The use of functionally graded materials to eliminate or control thermal deformation. *Compos. Sci. Technol.*, 56-1099.

[24] Tanigawa, Y., Morishita, H., & Ogaki, S. (1999). Derivation of system of fundamental equations for a three dimensional thermoelastic field with nonhomogeneous material properties and its application to a semi infinite body. *J. Therm. Stresses*, 22-689.

[25] Brush, D. O., & Almroth, B. O. (1975). Buckling of Bars Plates and Shells,. *McGraw-Hill, New York.*

[26] Reddy, J. N. (1999). Theory and Analysis of Elastic Plates. *Taylor and Francis, Philadelphia.*

[27] Song, G., Sethi, V., & Lic, H. N. (2006). Vibration control of civil structures using piezoceramic smart materials: A review. *Engineering Structures*, 28-1513.

[28] Efraim, E., & Eisenberger, M. (2007). Exact vibration analysis of variable thickness thick annular isotropic and FGM plate,. *J. Sound Vib.*, 299-720.

[29] Tzou, H. S. (1993). Piezoelectric Shells-(Distributed Sensing and Control of Continua,. *Kluwer Dordrecht.*

[30] Zheng, X. J., & Zhou, Y. H. (1990). Analytical formulas of solutions of geometrically nonlinear equations of axisymmetric plates and shallow shells. *Acta Mech. Sinica*, 6(1), 69-81.

[31] William, H. P., Brain, P. F., & Sau, A. T. (1986). Numerical Recipes-the Art of Scientific Computing,. *Cambridge University Press, New York.*

[32] Nayfe, H., & Mook, D. T. (1979). Nonlinear Oscillations. *John Wiley, New York.*

Chapter 2

Acoustic Wave Velocity Measurement on Piezoelectric Ceramics

Toshio Ogawa

Additional information is available at the end of the chapter

http://dx.doi.org/10.5772/53117

1. Introduction

1.1. Why is "acoustic wave velocity measurement" needed?

1.1.1. Relationships between Young's modulus and electromechanical coupling factor in piezoelectric ceramics and single crystals

Material research on lead-free piezoelectric ceramics has received much attention because of global environmental considerations. The key practical issue is the difficulty in realizing excellent piezoelectricity, such as electromechanical coupling factors and piezoelectric strain constants, by comparison with lead zirconate titanate [$Pb(Zr,Ti)O_3$] ceramics. The coupling factor is closely related to the degree of ferroelectric-domain alignment in the DC poling process. We have already shown the mechanism of domain alignment in $Pb(Zr,Ti)O_3$ (PZT) [1-5], $PbTiO_3$ (PT) [6], and $BaTiO_3$ (BT) [7] ceramics and in relaxor single crystals of $Pb[(Zn_{1/3}Nb_{2/3})_{0.91}Ti_{0.09}]O_3$ (PZNT91/09) [8] and $Pb[(Mg_{1/3}Nb_{2/3})_{0.74}Ti_{0.26}]O_3$ (PMNT74/26) [9] by measuring the piezoelectricity vs DC poling field characteristics. Throughout the studies, we have discovered relatively high coupling factor of transverse vibration mode k_{31} over 80 % (it called giant k_{31}) in the relaxor single-crystal plates [10,11]. Furthermore, we reported that the origin of giant k_{31} is due to the lowest Young's modulus of the single-crystal plates [11,12]. The results indicate the important relationships between Young's modulus and electromechanical coupling factor to realize high piezoelectricity.

Concrete experimental results are mentioned as follows. The Young's modulus (Y^E) of PZNT91/09 single crystals with giant k_{31} (Y^E=0.89x10^{10} N/m^2) is one order of magnitude smaller than the Y^E (6~9x10^{10} N/m^2) of PZT ceramics, and roughly speaking, one order of magnitude larger than the Y^E (0.05x10^{10} N/m^2) of rubber [12]. It was thought that the origin of giant k_{31} in PZNT91/09 and PMNT74/26 single crystals was due to the mechanical softness of the materials. Namely, the most important factor to realize high piezoelectricity is easy deformation by the DC poling field. Figure 1 shows the relationships with k_{31} and k_{33} (a coupling factor of longitudinal vibration mode) vs Y^E on various kinds of single crystals and ceramics reported [13]. There is a linear relationship with k_{31} vs Y^E and k_{33} vs Y^E. Furthermore, there is a blank space between Y^E of PZNT91/09 and PMNT74/26 single crystals and Y^E of ordinary piezoelectric materials. Therefore, the important viewpoint is that new piezoelectric materials including lead-free compositions with higher coupling factor should be investigated to clarify the blank space, such as the research for elements softened the materials in lead-free ceramics.

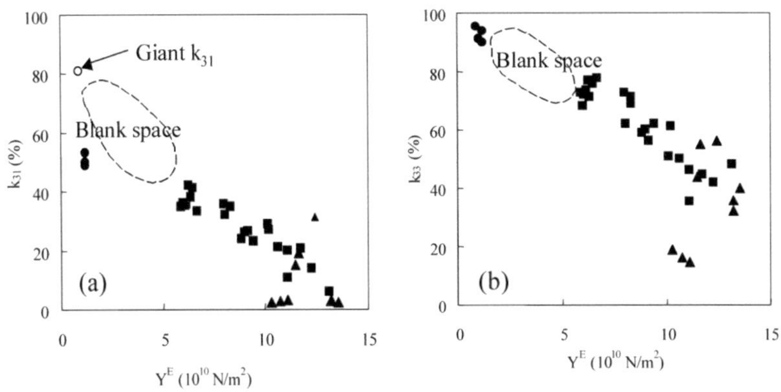

Figure 1. Relationships with coupling factors of (a) k_{31} and (b) k_{33} vs Young's modulus (Y^E) in piezoelectric materials.

1.1.2. Newly developed measurement method for acoustic wave velocities

Elastic constants of piezoelectric ceramics such as Young's modulus and Passion's ratio were basically evaluated by impedance responses on frequency of piezoelectric resonators with various relationships between DC poling directions and vibration modes [14-17]. Therefore, it needs all kinds of resonators with different shapes [14], and furthermore, with uniform poling-degree in spite of different poling-thicknesses such as 1.0 mm for plate resonators (for example, dimensions of 12 mm length, 3.0 mm width and 1.0 mm poling-thickness) and 15 mm for rod resonators (dimensions of 6.0 mm diameter and 15 mm poling-thickness). While applying a DC poling field of 3.0 kV/mm, 3.0 kV DC voltage must be applied to the former sample and 45 kV for the later sample. There was no guarantee to obtain uniform poling-degree between two samples even through the same poling field of 3.0 kV/mm. In addition, it is difficult to measure in the cases of as-fired ceramics, namely ceramics before poling, or weak polarized ceramics because of none or weak impedance responses on frequency.

On the other hand, there was a well-known method to measure the elastic constants by pulse echo measurement [18,19]. Acoustic wave velocities in ceramics were directly measured by this method. However, since the frequency of pulse oscillation for measurement was generally below 5 MHZ, it needs to prepare the rod samples with thickness of 10-20 mm in order to guarantee accuracy of the velocities. Moreover, it is unsuitable for measuring samples such as disks with ordinary dimensions (10-15 mm diameter and 1.0-1.5 mm thickness) and with many different compositions for piezoelectric materials R & D.

We developed a method to be easy to measure acoustic wave velocities suitable for the above mentioned disk samples by an ultrasonic thickness gauge with high-frequency pulse oscillation. This method was applied to hard and soft PZT [3], lead titanate [6] (both lead-containing ceramics) and lead-free ceramics composed of alkali niobate [20,21] and alkali bismuth titanate [22]. In this chapter, first of all, we report on the measurement of acoustic wave velocities in PZT ceramics, and the following, the calculation results of Young's modulus and Poisson's ratio, especially to obtain high piezoelectricity in lead-free ceramics.

For the measurement, the ultrasonic precision thickness gauge (Olympus Co., Model 35DL) has PZT transducers with 30 MHz for longitudinal-wave generation and 20 MHz for transverse-wave generation. The acoustic wave velocities were evaluated by the propagation time between second pulse echoes in thickness of ceramic disks with dimensions of 14 mm in diameter and 0.5-1.5 mm in thickness. The sample thickness was measured by a precision micrometer (Mitutoyo Co., Model MDE-25PJ). Piezoelectric ceramic compositions measured were as follows: $0.05Pb(Sn_{0.5}Sb_{0.5})O_3$-(0.95-x)$PbTiO_3$-x$PbZrO_3$ (x=0.33, 0.45, 0.48, 0.66, 0.75) with (hard PZT) and without 0.4 wt% MnO_2 (soft PZT) [3]; (1-x)(Na,K,Li,Ba)$(Nb_{0.9}Ta_{0.1})O_3$-x$SrZrO_3$(SZ) (x=0, 0.02, 0.04, 0.05, 0.06, 0.07) [20,21]; (1-x)$(Na_{0.5}Bi_{0.5})TiO_3$(NBT)-x$(K_{0.5}Bi_{0.5})TiO_3$ (KBT) (x=0.08, 0.11, 0.15, 0.18, 0.21, 0.28) and 0.79NBT-0.20KBT-0.01$Bi(Fe_{0.5}Ti_{10.5})O_3$(BFT) [22]; and (1-x)NBT-x$BaTiO_3$(BT) (x=0.03, 0.07, 0.11) [22]. In addition to evaluate ceramic compositions, we investigated ceramics with different ceramic manufacturing processes such as firing processes of normal firing in air atmosphere and oxygen atmosphere firing to realize pore-free ceramics [23], and DC poling processes of as-fired (before poling), weakly and fully polarized ceramics.

2. Young's modulus and Poisson's ratio in piezoelectric ceramics

2.1. Longitudinal and transverse wave velocities by pulse echo measurement

Figure 2 shows pulse echoes of longitudinal acoustic wave in hard PZT ceramics at a composition of x=0.48. The longitudinal wave velocity was calculated from the propagation time between second-pulse echoes (␣) and the thickness of ceramic disks. The dependences of the longitudinal and transverse wave velocities on the composition of x in hard and soft PZT ceramics and their fluctuation of the individual ceramic disks were shown in Figs. 3(a)-(d). The fluctuation of the velocities in transverse wave was smaller than the ones in longitudinal wave when the samples of n=16-21 pieces were measured at each composition. In addition, it was clarified that the fluctuation in soft PZT was smaller than the one in hard PZT ceramics because of easy alignment of ferroelectric domains by poling field in soft PZT, namely lower coercive fields than the ones of hard PZT.

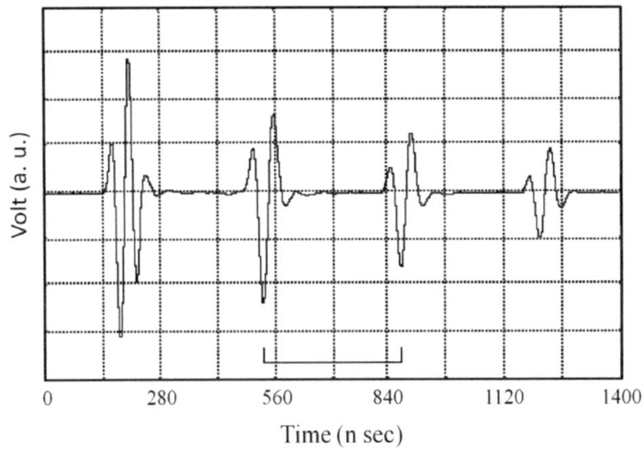

Figure 2. Pulse echoes of longitudinal acoustic wave in hard PZT ceramics (disk dimensions: 13.66 mm diameter and 0.735 mm thickness) before poling at x=0.48; the calculated wave velocity is 4,319 m/s.

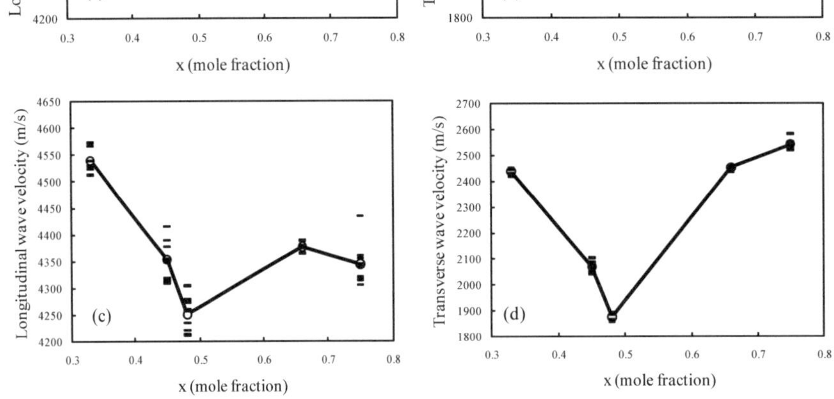

Figure 3. Composition x dependence of longitudinal and transverse wave velocities in hard [(a) and (b)] and soft PZT [(c) and (d)] ceramics before poling; all the samples of n=16-21 are shown in the figure to evaluate their fluctuation.

2.2. Young's modulus and Poisson's ratio in PZT ceramics

Figures 4(a)-(d) show the dependences of the longitudinal and transverse wave velocities, Young's modulus and Poisson's ratio on the composition of x in hard and soft PZT ceramics, respectively. Here, the elastic constants were calculated from equations in the following session 3.1. From Fig. 4, it was found that the large differences in these values between hard and soft PZT ceramics only occurred around a composition of x=0.48, which corresponds to a morphotropic phase boundary (MPB) [23].

In the case of the evaluation on ceramic manufacturing processes, the values of pore-free ceramics fired in oxygen atmosphere also show in the figures at compositions of x=0.66 and 0.75 (in dotted circles, the samples of n=2). The effect of reducing pores in ceramics on the values works like the increase of the longitudinal wave velocity and Poisson's ratio, and the decrease of the transverse wave velocity and Young's modulus.

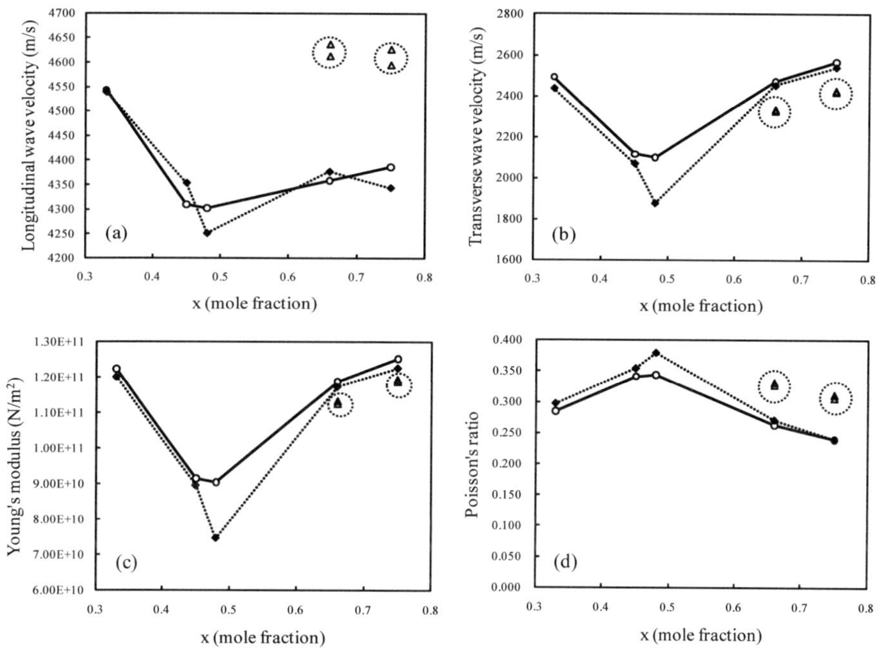

Figure 4. Composition x dependence of (a) longitudinal and (b) transverse wave velocities, (c) Young's modulus and (d) Poisson's ratio in hard (—) and soft (---) PZT ceramics before poling; the values of pore-free hard PZT ceramics (x=0.66, 0.75 and the samples of n=2) were shown in dotted circles.

As mentioned later in the session 4, the decrease in Young's modulus and the increase in Poisson's ratio correspond to improve the piezoelectricity. Therefore, it was confirmed by using the pulse echo measurement that oxygen atmosphere firing is an effective tool to improve piezoelectric properties as well as production of pore-free ceramics. Figures 5(a)-(d) show the dependences of the longitudinal and transverse wave velocities, Young's modulus and Poisson's ratio on the firing position in sagger at a composition of x=0.45 in hard PZT ceramics, respectively. Nos. 1 and 10 in the figure correspond to the top position piled up and the bottom position piled up in sagger (Fig. 6). It became clear that there was fluctuation of the velocities and elastic constants in the case of as-fired (before poling) samples. In addition, the sample of the middle position piled up of No. 5 possesses low wave velocities, low Young's modulus and high Poisson's ratio because of the firing under high lead oxide (PbO) atmosphere. On the other hand, the samples of the top and bottom positions of Nos. 1, 2 and 10 possess high wave velocities, high Young's modulus and low Poisson's ratio because of the firing under low PbO atmosphere due to vaporization of PbO during firing (see Fig. 6). Young's modulus and Poisson's ratio in PZT ceramics of No. 5 could be improved the piezoelectricity (see the session of 4) by high PbO atmosphere firing. However, there is a different tendency to decrease the longitudinal wave velocity [No.5 position in Fig. 5(a)] in comparison with the case of oxygen atmosphere firing [dotted circles at x=0.66, 0.75 in Fig. 4(a)], which also can be improved the elastic constants.

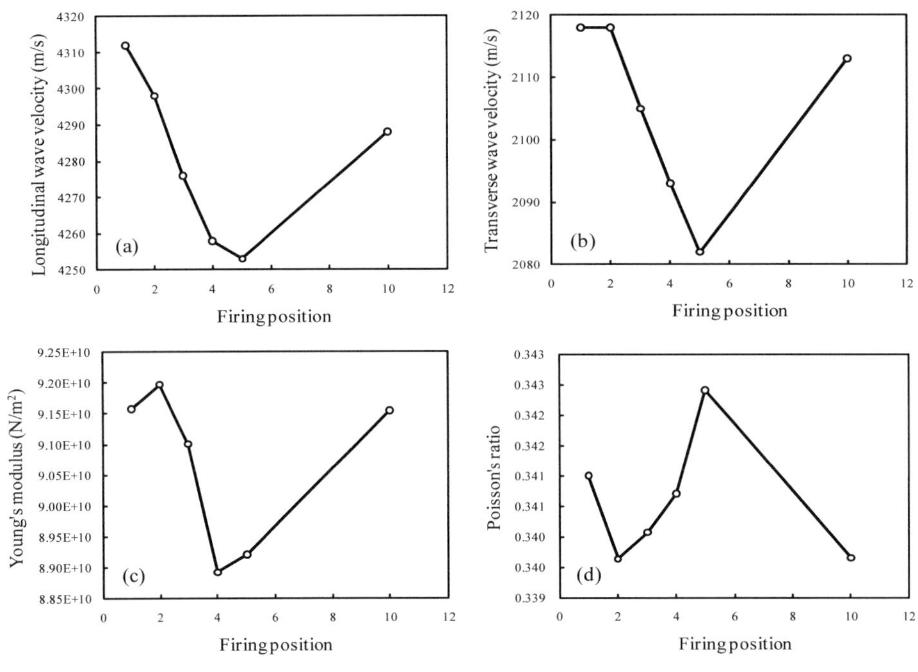

Figure 5. Firing position (Nos. 1 and 10 correspond to ceramic disks of top piled up and bottom piled up in sagger) dependence of (a) longitudinal and (b) transverse wave velocities, (c) Young's modulus and (d) Poisson's ratio in hard PZT ceramics at x=0.45.

We believe that the increase in longitudinal wave velocity [dotted circles at x=0.66, 0.75 in Fig. 4(a)] comes from pore-free microstructure, and further, higher dense and higher Passion's ratio ceramics compared with in the case of normal firing in air atmosphere. Furthermore, it is thought that the fluctuation of as-fired ceramics connected to the fluctuation of ceramics after DC poling.

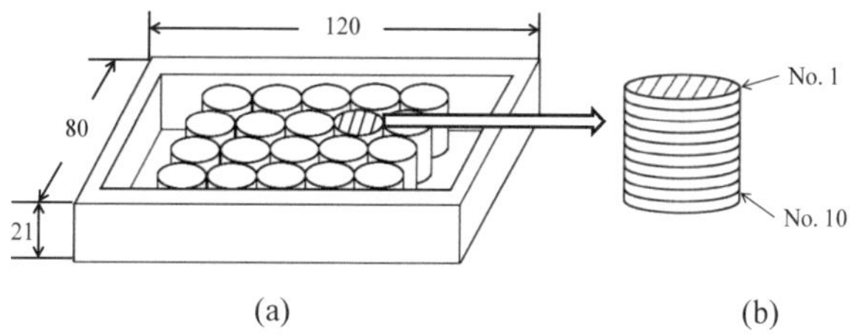

Figure 6. Schematic pictures of (a) PZT green (as-pressed) disks piled up in dense Al_2O_3 sagger (container dimensions: 120 mm length, 80 mm width and 21 mm height) and (b) firing positions (Nos. 1-10) of ceramic green disks (16 mm diameter and 1.2 mm thickness); the ceramic green disk of No. 5 is surrounded with high PbO atmosphere by the other disks piled up, and it is understood that PbO vaporization is easy to occur at the disks of Nos.1, 2 (top) and No.10 (bottom) even through a cover (dense Al_2O_3 ceramic plate) is put on the container while firing [23].

2.3. Young's modulus and Poisson's ratio in lead-free ceramics

Figures 7(a)-(d) show the dependences of the longitudinal and transverse wave velocities, Young's modulus and Poisson's ratio on the composition of lead-free ceramics after poling; the compositions were (1-x)(Na,K,Li,Ba)($Nb_{0.9}Ta_{0.1}$)O_3-xSZ (x=0, 0.02, 0.04, 0.05, 0.06, 0.07), (1-x)NBT-xKBT (x=0.08, 0.11, 0.15, 0.18, 0.21, 0.28), 0.79NBT-0.20KBT-0.01BFT and (1-x)NBT-xBT (x=0.03, 0.07, 0.11), respectively. While the compositions to obtain high electromechanical coupling factor in (Na,K,Li,Ba)($Nb_{0.9}Ta_{0.1}$)O_3-SZ (x=0.05-0.06) and NBT-KBT (x=0.18) [21,22] correspond to the compositions with low Young's modulus and high Poisson's ratio as well as PZT (x=0.48) [3]. Although a MPB existed around x=0.18 in the phase diagram of NBT-KBT [24,25], a MPB did not exist in the phase diagram of (Na,K,Li,Ba)($Nb_{0.9}Ta_{0.1}$)O_3-SZ [20,21]. On the other hand, the compositions with high coupling factor in NBT-BT (x=0.07) [22] possessed low Young's modulus and low Poisson's ratio such as $PbTiO_3$ [6]. In the phase diagram of NBT-BT, it is confirmed that there is a MPB near x=0.07 [26]. Furthermore, there was no significant difference in Young's modulus and Poisson's ratio between as-fired (△) (before poling) and after poling (▲) in (Na,K,Li,Ba)($Nb_{0.9}Ta_{0.1}$)O_3-SZ.

From the above mentioned viewpoint of the study on the elastic constants in lead-free ceramic, it was clarified that there are compositions with high coupling factor in the cases of (1) low Young's modulus and high Poisson's ratio (PZT type) and (2) low Young's modulus and low Poisson's ratio ($PbTiO_3$ type). Namely, it was found that there are two kinds of MPB compositions, (1) and (2) in lead-free ceramics with high coupling factor. Therefore, lead-free ceramic compositions with high piezoelectricity must be primarily focused on to realize the compositions with low Young's modulus such as MPB compositions [27]. Secondarily, the importance to measure Poisson's ratio was understood for the research to recognize the compositions of (1) and (2) in lead-free ceramics with high piezoelectricity.

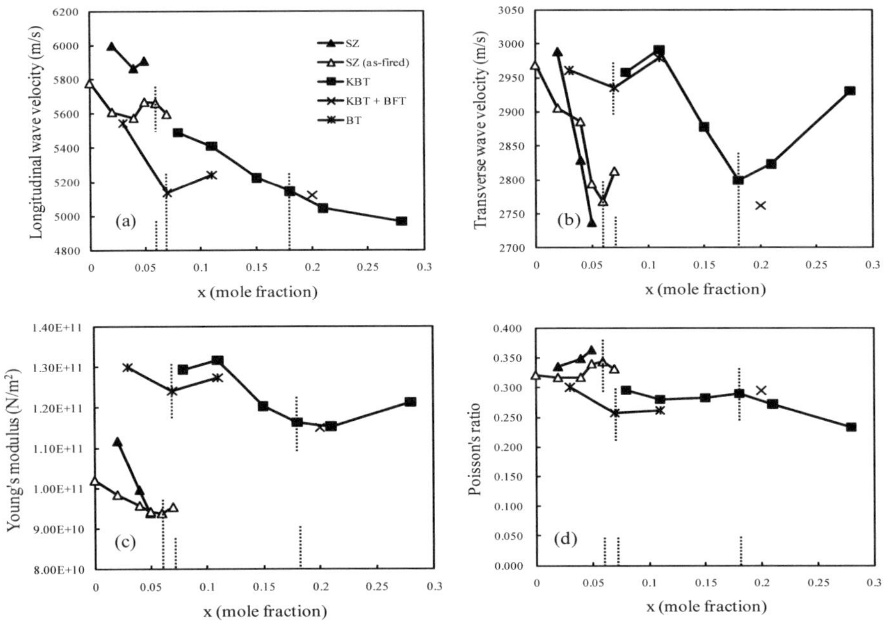

Figure 7. Composition x dependence of (a) longitudinal and (b) transverse wave velocities, (c) Young's modulus and (d) Poisson's ratio in lead-free ceramics of (1-x)(Na,K,Li,Ba)($Nb_{0.9}Ta_{0.1}$)O_3-xSZ (x=0, 0.02, 0.04, 0.05, 0.06, 0.07), (1-x)NBT-xKBT (x=0.08, 0.11, 0.15, 0.18, 0.21, 0.28), 0.79NBT-0.20KBT-0.01BFT and (1-x)NBT-xBT (x=0.03, 0.07, 0.11).

3. Poling field dependence of longitudinal and transverse wave velocities, Young's modulus and Poisson's ratio in piezoelectric ceramics

3.1. DC poling field dependence of elastic constants

As mentioned previously in the session 1, the behavior of ferroelectric domains toward a DC field was investigated on the basis of the DC poling field dependence of dielectric and piezoelectric properties in various types of piezoelectric ceramics and single crystals. In addition, we have developed a method to evaluate elastic constants, such as Young's modulus and Passion's ratio, by measuring longitudinal and transverse wave velocities using an ultrasonic thickness gauge with high-frequency pulse oscillation in comparison with a conventional method as described in the session 2. The acoustic wave velocities can be measured by this method in the cases of ceramics and single crystals despite the strength of the DC poling field, including as-fired ceramics and as-grown single crystals.

Therefore, in order to clarify the relationships between elastic constants and electrical properties vs DC poling fields, we studied the poling field dependence of acoustic wave velocities and dielectric and piezoelectric properties in ceramics. Here, we report the relationships between DC poling fields, acoustic wave velocities, Young's modulus, and Passion's ratio to realize a high piezoelectricity, especially in lead-free ceramics.

The piezoelectric ceramic compositions measured were as follows: $0.05Pb(Sn_{0.5}Sb_{0.5})O_3$-(0.95-x)$PbTiO_3$-x$PbZrO_3$ (x=0.33, 0.45, 0.66) with (hard PZT) and without 0.4 wt% MnO_2 (soft PZT); $0.90PbTiO_3$-$0.10La_{2/3}TiO_3$(PLT) and $0.975PbTiO_3$-$0.025La_{2/3}TiO_3$(PT); (1-x) (Na,K,Li,Ba)($Nb_{0.9}Ta_{0.1}$) O_3-x$SrZrO_3$(SZ) (x=0.00, 0.05, 0.07); (1-x)($Na_{0.5}Bi_{0.5}$)TiO_3(NBT)-x($K_{0.5}Bi_{0.5}$)TiO_3(KBT) (x=0.08, 0.18) and 0.79NBT-0.20KBT-0.01Bi($Fe_{0.5}Ti_{10.5}$)O_3(BFT) (x= 0.20); and (1-x)NBT-x$BaTiO_3$(BT) (x=0.03, 0.07, 0.11).

DC poling was conducted for 30 min at the most suitable poling temperature depending on the Curie points of the ceramic materials when the poling field (E) was varied from 0→ +0.25→+0.5→+0.75→ +1.0→ --- $+E_{max}$→0→$-E_{max}$→0 to $+E_{max}$ kV/mm. $\pm E_{max}$ depended on the coercive fields of the piezoelectric ceramics. After each poling process, the dielectric and piezoelectric properties were measured at room temperature using an LCR meter (HP4263A), a precision impedance analyzer (Agilent 4294A), and a piezo d_{33} meter (Academia Sinica ZJ-3D). Furthermore, the acoustic wave velocities were measured using an ultrasonic precision thickness gauge (Olympus 35DL), which has PZT transducers with 30 MHz for longitudinal-wave generation and 20 MHz for transverse-wave generation. The acoustic wave velocities were evaluated on the basis of the propagation time between the second-pulse echoes in the thickness of ceramic disks parallel to the poling field with dimensions of 14 mm diameter and 0.5-1.5 mm thickness. The sample thickness was measured using a precision micrometer (Mitutoyo MDE-25PJ). Moreover, Young's modulus in the thickness direction of ceramic disks ($Y_{33}{}^E$) and Passion's ratio (σ) were calculated on the basis of the longitudinal (V_L) and transverse (V_S) wave velocities as shown in the following equations:

$$Y_{33}{}^{E}=3\varrho V_S{}^2\frac{V_L{}^2-\frac{4}{3}V_S{}^2}{V_L{}^2-V_S{}^2} \text{ and } \sigma=\frac{1}{2}\left\{1-\frac{1}{\left(\frac{V_L}{V_S}\right)^2-1}\right\},$$

where ϱ is the density of ceramic disks.

3.2. Poling field dependence in PZT ceramics

Figures 8-10 show the poling field dependence of longitudinal (V_L) and transverse (V_S) wave velocities, Young's modulus ($Y_{33}{}^E$), and Passion's ratio (σ) in $0.05Pb(Sn_{0.5}Sb_{0.5})O_3$-(0.95-x)$PbTiO_3$-x$PbZrO_3$ (x=0.33, 0.45, 0.66) with (hard PZT) and without 0.4 wt% MnO_2 (soft PZT) ceramics at a poling temperature (T_P) of 80 °C (hard PZT in Fig. 8 and soft PZT in Fig. 9), $0.90PbTiO_3$-$0.10La_{2/3}TiO_3$ (abbreviate to PLT/T_p=80 °C) and $0.975PbTiO_3$-$0.025La_{2/3}TiO_3$ (abbreviate to PLT /T_p=200 °C) ceramics (Fig. 10), respectively. While the poling field dependence of V_L has almost same tendency in spite of hard and soft PZT, the one of V_S at x=0.45, which corresponds to MPB, abruptly decreases in the both cases of hard and soft PZT (Figs. 8, 9). Since the highest coupling factor in PZT is obtained at the MPB (x=0.45), the origin of the highest piezoelectricity is due to the decrease in V_S with increasing the domain alignment by DC poling field. Furthermore, the lowest $Y_{33}{}^E$ and the highest σ are realized at the MPB. The change in V_L, V_S, $Y_{33}{}^E$ and σ vs E of soft PZT is smaller than the one of hard PZT because of the softness of the materials. As mentioned details to the next session 3.3, it was indicated that minimum V_L, σ and maximum V_S, Y_{33}^E were obtained at the domain clamping such as the domain alignment canceled each other (↑↓), at which the lowest piezoelectricity is realized. On the other hand, the values of V_L, V_S, Y_{33}^E and σ vs E in PT are smaller than the ones of PLT. Moreover, at the domain clamping fields, minimum V_L, σ, Y_{33}^E and maximum V_S were obtained in both the PLT and PT. The reason of minimum Y_{33}^E at the DC field of the domain clamping will be discuss in the next session. In addition, the change in Y_{33}^E of PLT and PT while applying ±E is smaller than the one of PZT, and higher Y_{33}^E and lower σ appear in comparison with the ones of PZT, because these come from the hardness of PLT and PT ceramics.

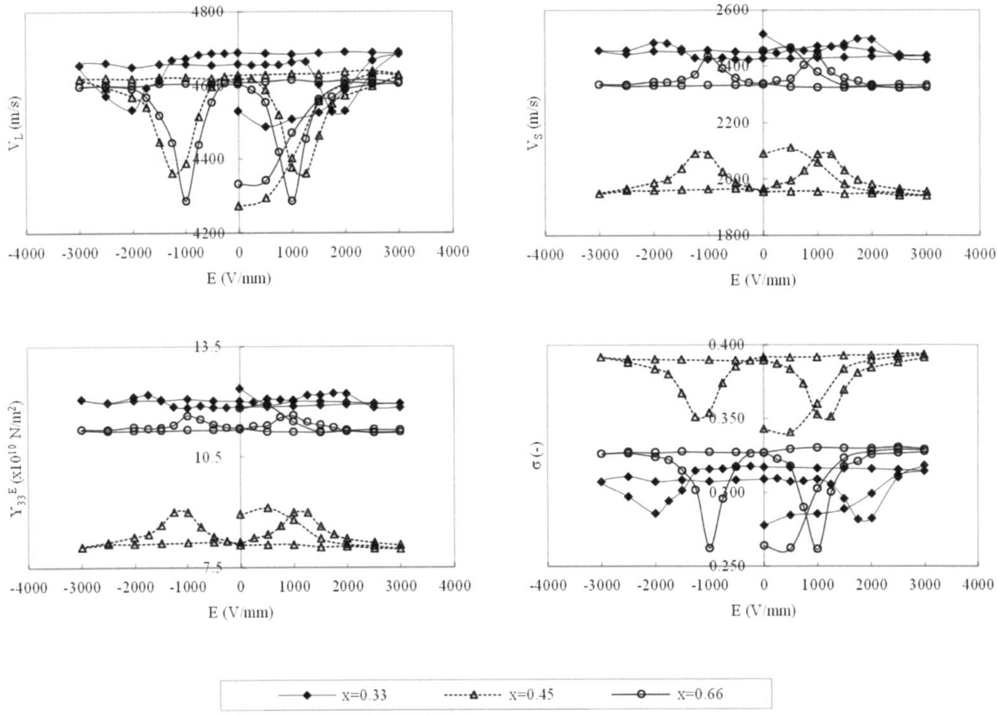

Figure 8. Poling field dependence of longitudinal (V_L) and transverse (V_S) wave velocities, Young's modulus (Y_{33}^E), and Poisson's ratio (σ) in $0.05Pb(Sn_{0.5}Sb_{0.5})O_3$-$(0.95\text{-}x)PbTiO_3$-$xPbZrO_3$ (x=0.33, 0.45, 0.66) with 0.4 wt% MnO_2 (hard PZT) ceramics.

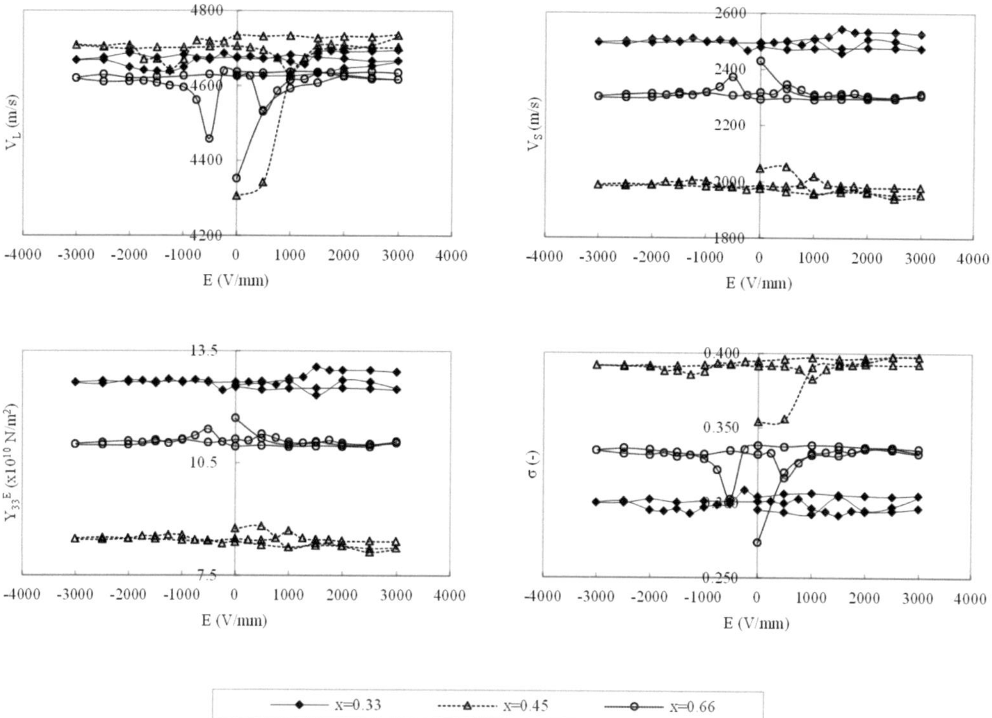

Figure 9. Poling field dependence of longitudinal (V_L) and transverse (V_S) wave velocities, Young's modulus (Y_{33}^E), and Poisson's ratio (σ) in $0.05Pb(Sn_{0.5}Sb_{0.5})O_3$-(0.95-x)$PbTiO_3$-x$PbZrO_3$ (x=0.33, 0.45, 0.66) without 0.4 wt% MnO_2 (soft PZT) ceramics.

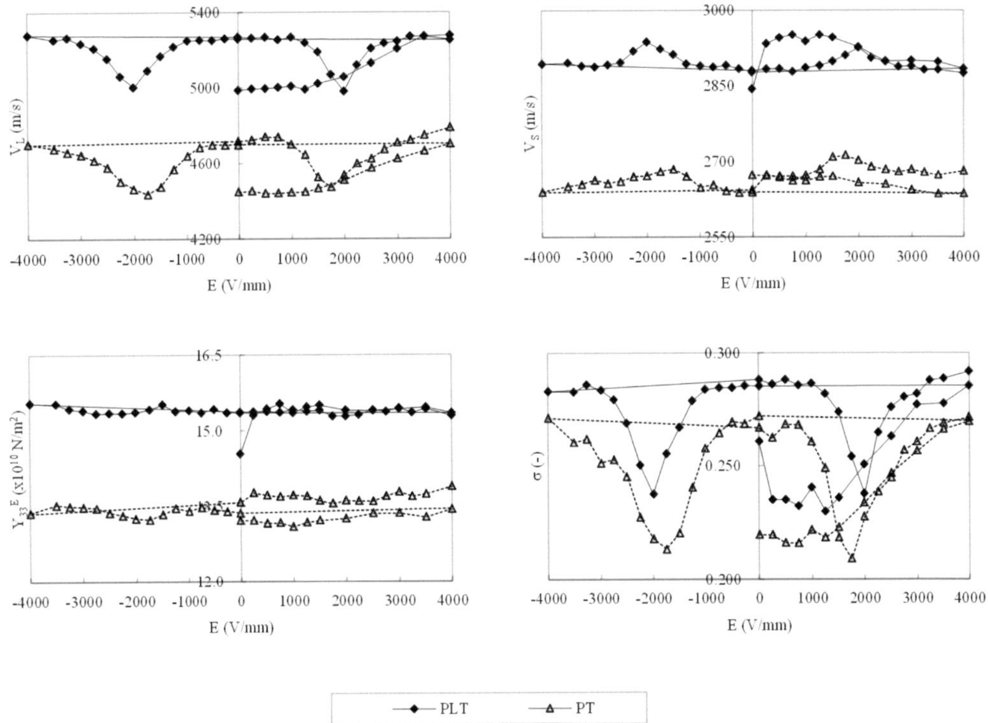

Figure 10. Poling field dependence of longitudinal (V_L) and transverse (V_S) wave velocities, Young's modulus (Y_{33}^E), and Poisson's ratio (σ) in 0.90PbTiO$_3$-0.10La$_{2/3}$TiO$_3$ (PLT) and 0.975PbTiO$_3$-0.025La$_{2/3}$TiO$_3$ (PLT) ceramics.

3.3. Poling field dependence in lead-free ceramics

Figures 11-13 show the poling field dependence of longitudinal (V_L) and transverse (V_S) wave velocities, Young's modulus (Y_{33}^E), and Passion's ratio (σ) in (1-x)(Na,K,Li,Ba)(Nb$_{0.9}$Ta$_{0.1}$)O$_3$-xSrZrO$_3$(SZ) (x=0.00, 0.05, 0.07) ceramics at a poling temperature (T_P) of 150 °C (Fig. 11), in (1-x)(Na$_{0.5}$Bi$_{0.5}$)TiO$_3$(NBT)-x(K$_{0.5}$Bi$_{0.5}$)TiO$_3$(KBT) (x=0.08, 0.18/ T_p=70 °C) and 0.79NBT-0.20KBT-0.01Bi(Fe$_{0.5}$Ti$_{10.5}$)O$_3$(BFT) (x=0.20/ T_p=70 °C) ceramics (Fig. 12), and in (1-x)NBT-xBaTiO$_3$(BT) (x=0.03, 0.07, 0.11/ T_p=70 °C) ceramics (Fig. 13), respectively. With the enhancement of domain alignment with an increase in poling field from E=0 to +E_{max}, V_L increased and V_S decreased independently of the ceramic composition. From the composition dependence of Y_{33}^E and σ, high piezoelectricity [high planar coupling factor (k_p)] compositions show lower Y_{33}^E and higher σ values at 0.95(Na,K,Li,Ba)(Nb$_{0.9}$Ta$_{0.1}$)O$_3$-0.05SZ (k_p=46%) (Fig. 11) and 0.82NBT-0.18KBT (k_p=27%) (Fig. 12) as well as the ones at 0.05Pb(Sn$_{0.5}$Sb$_{0.5}$)O$_3$-0.47PbTiO$_3$-0.48PbZrO$_3$ (k_p=65% in soft ceramics and k_p=52% in hard ceramics) than the other compositions. Although morphotropic phase boundaries (MPBs) were observed in the NBT-KBT [24,25] and PZT [3,23,28] ceramics, there was no evidence of the existence of MPBs in the (Na,K,Li,Ba)(Nb$_{0.9}$Ta$_{0.1}$)O$_3$-SZ ceramics [20,21]. The effects of 0.01BFT modification in NBT-KBT on Y_{33}^E and σ were as follows: the composition of

http://dx.doi.org/10.5772/53117

0.79NBT-0.20KBT-0.01BFT (k_p=22%; x=0.20 in Fig. 12) showed the highest piezoelectric d_{33} constant of 150 pC/N in NBT-KBT and higher Y_{33}^E and lower σ values than that of 0.82NBT-0.18KBT (k_p=27%; x=0.18 in Fig. 12) because the relative dielectric constant increased from 800 (0.82NBT-0.18KBT) to 1250 (0.79NBT-0.20KBT-0.01BFT) [22]. Moreover, a high-coupling-factor composition at 0.93NBT-0.07BT (k_p=16%) existed in a MPB [26] with a low Y_{33}^E (Fig. 13).

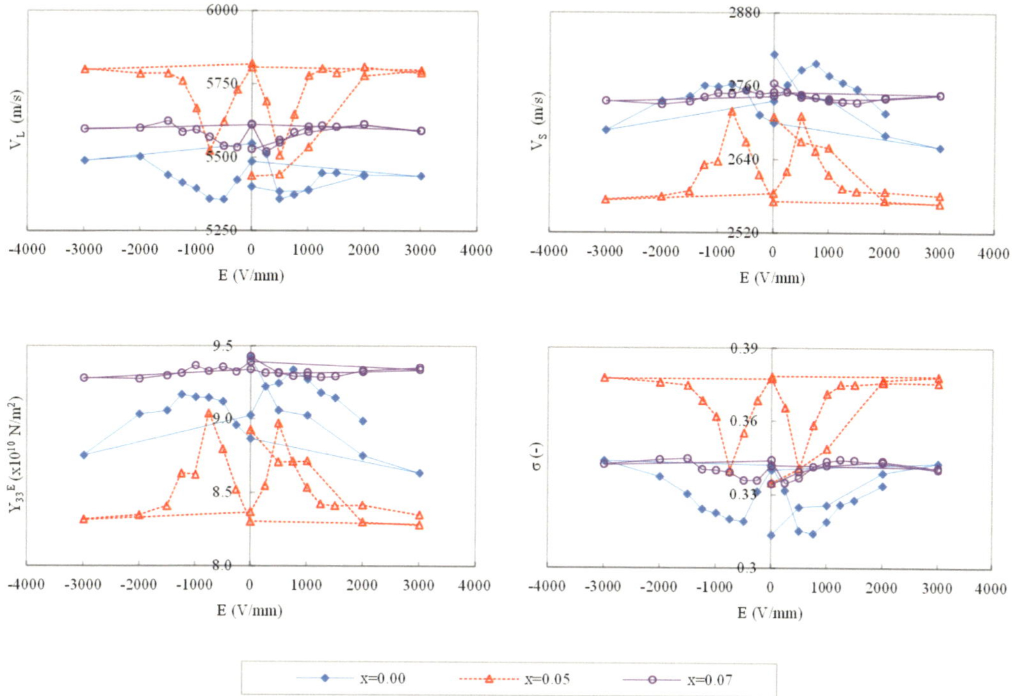

Figure 11. Poling field dependence of longitudinal (V_L) and transverse (V_S) wave velocities, Young's modulus (Y_{33}^E), and Poisson's ratio (σ) in (1-x)(Na,K,Li,Ba)($Nb_{0.9}Ta_{0.1}$)O_3-x$SrZrO_3$ (x=0.00, 0.05, 0.07) ceramics.

Basically, research on piezoelectric ceramics with high planar coupling factors, especially in lead-free ceramics, has been focused on determining the MPB composition because many different polarization axes are generated in MPB.

Through our study on acoustic wave measurement, it was found that there was an important factor for obtaining a high piezoelectricity regarding Y_{33}^E and σ; lower Y_{33}^E and higher σ

values in the cases of PZT, alkali bismuth titanate [NBT-KBT], and alkali bismuth barium titanate [NBT-BT] with MPB, and in the case of alkali niobate [$(Na,K,Li,Ba)(Nb_{0.9}Ta_{0.1})O_3$-SZ] without MPB. Therefore, as mentioned later, we need a new concept in addition to conventional research on MPB compositions for developing piezoelectric ceramics with high coupling factors.

The DC poling fields under domain clamping (E_d) can be estimated on the basis of E values to realize the minimum V_L and maximum V_S in Figs. 11-13. Although the maximum Y_{33}^E and minimum σ were obtained at E_d, which corresponds to the E of the minimum coupling factor in $(Na,K,Li,Ba)(Nb_{0.9}Ta_{0.1})O_3$-SZ (Fig. 11), NBT-KBT, and 0.79NBT-0.20KBT-0.01BFT (Fig. 12), the minimum Y_{33}^E and minimum σ were obtained at E_d in NBT-BT (Fig. 13). It was clarified that the minimum Y_{33}^E in NBT-BT was realized in the cases of $\Delta V_S/\Delta V_L<1/4$ at E_d, where ΔV_S and ΔV_L denote the variations in V_S and V_L at E_d, respectively. This may be due to the poor domain alignment (lower Poisson's ratio) perpendicular to the poling field (radial direction in disk ceramics) in comparison with the domain alignment parallel to the poling field (thickness direction in disk ceramic). Furthermore, the increase in coupling factor corresponds to the increase in σ at all compositions including lead-containing and lead-free ceramic compositions.

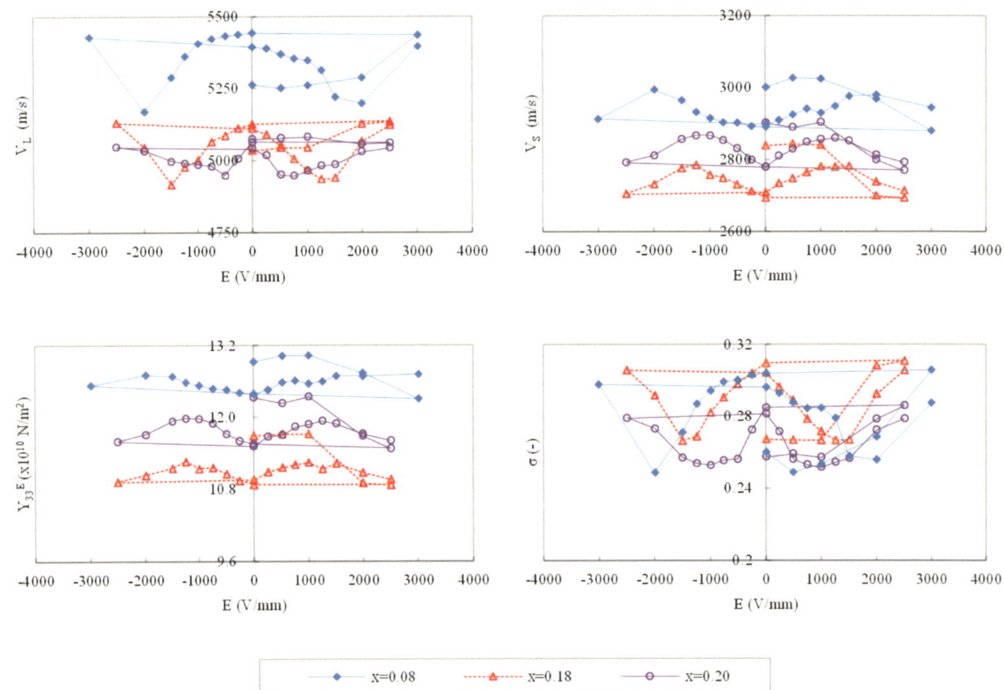

Figure 12. Poling field dependence of longitudinal (V_L) and transverse (V_S) wave velocities, Young's modulus (Y_{33}^E), and Poisson's ratio (σ) in (1-x)NBT-xKBT (x=0.08, 0.18) and 0.79NBT-0.2KBT-0.01BFT (x=0.20) ceramics.

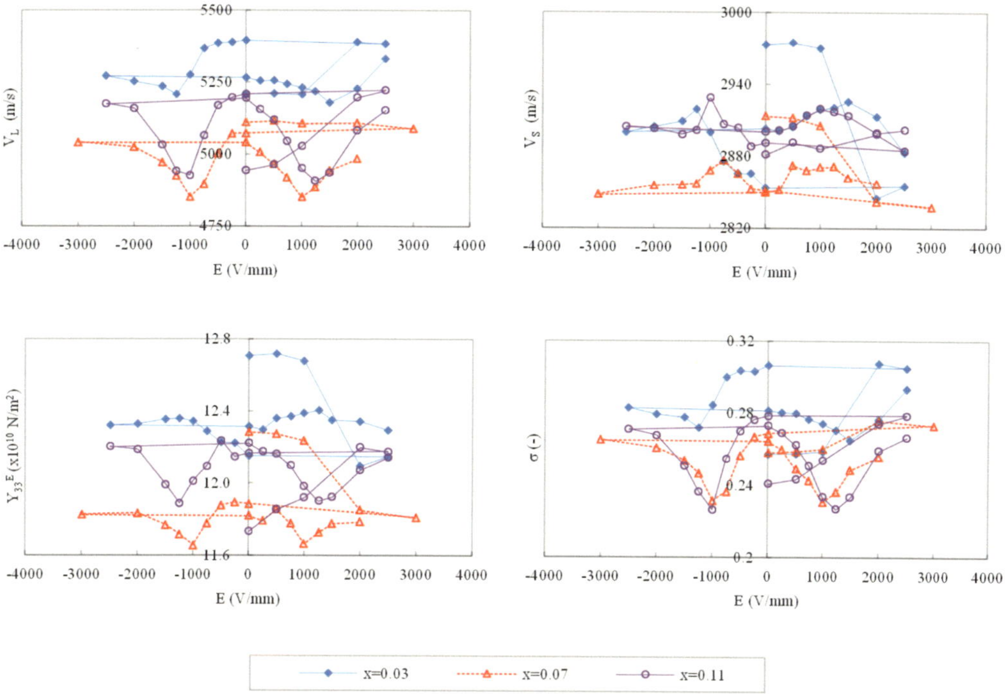

Figure 13. Poling field dependence of longitudinal (V_L) and transverse (V_S) wave velocities, Young's modulus (Y_{33}^E), and Poisson's ratio (σ) in (1-x)NBT-xBT (x=0.03, 0.07, 0.11) ceramics.

4. Materials road maps in piezoelectric ceramics on elastic constants

Figure 14 shows the relationships between longitudinal (V_L) and transverse (V_S) wave velocities, Young's modulus (Y_{33}^E), and Passion's (σ) ratio vs planar coupling factors (k_p) in (1-x)(Na,K,Li,Ba)($Nb_{0.9}Ta_{0.1}$)O_3-xSZ (abbreviated to "SZ"), (1-x)NBT-xKBT ("KBT"), 0.79NBT-0.20KBT-0.01BFT ("KBT"), and (1-x)NBT-xBT ("BT") lead-free ceramics compared with 0.05Pb($Sn_{0.5}Sb_{0.5}$)O_3-(0.95-x)$PbTiO_3$-x$PbZrO_3$ ceramics with ("hard PZT") and without 0.4 wt % MnO_2 ("soft PZT"), and with 0.90$PbTiO_3$-0.10$La_{2/3}TiO_3$ ("PLT") and 0.975$PbTiO_3$-0.025 $La_{2/3}TiO_3$ ("PT") lead-containing ceramics after full DC poling. Although the V_L values of the PZT ceramics were almost constant at approximately 4,600-4,800 m/s independently of the composition x, their V_S values linearly decreased from 2,500 to 1,600 m/s with increasing k_p from 20 to 65% (solid lines). In addition, the V_L and V_S values of the PZT ceramics were smaller than those of the lead-free ceramics (V_L=5,000-5,800 m/s and V_S=2,600-3,000 m/s; dashed and dotted lines). Although the V_L values of the PT ceramics were almost the same (4,800 m/s) as those of the PZT ceramics, the V_S values of the PT ceramics were approximately 2,700 m/s. On the other hand, the V_L values of the SZ ceramics were relatively high (5,500-5,800 m/s); furthermore, the V_S values of the SZ ceramics also increased (2,600-2,700 m/s) and linearly decreased with increasing k_p from 25 to 50% (dashed lines), the behavior of which was almost the same as that of the V_S values of the PZT ceramics. The V_L values of the KBT, BT, and PLT ceramics (5,000-5,400 m/s) were between those of the PZT, PT, and SZ ceramics. However, the V_S values of the KBT, BT, and PLT ceramics (2,800-3,000 m/s) were the highest. Therefore, it was possible to divide V_L and V_S into three material groups, namely, PZT and PT/ KBT, BT (alkali bismuth titanate), and PLT/ SZ (alkali niobate). In addition, k_p lineally increased from 4 to 65% with decreasing Y_{33}^E from 15×10^{10} to 6×10^{10} N/m^2 and lineally increased with increasing σ from 0.25 to 0.43. It was clarified that higher k_p values can be realized at lower Y_{33}^E and higher σ values.

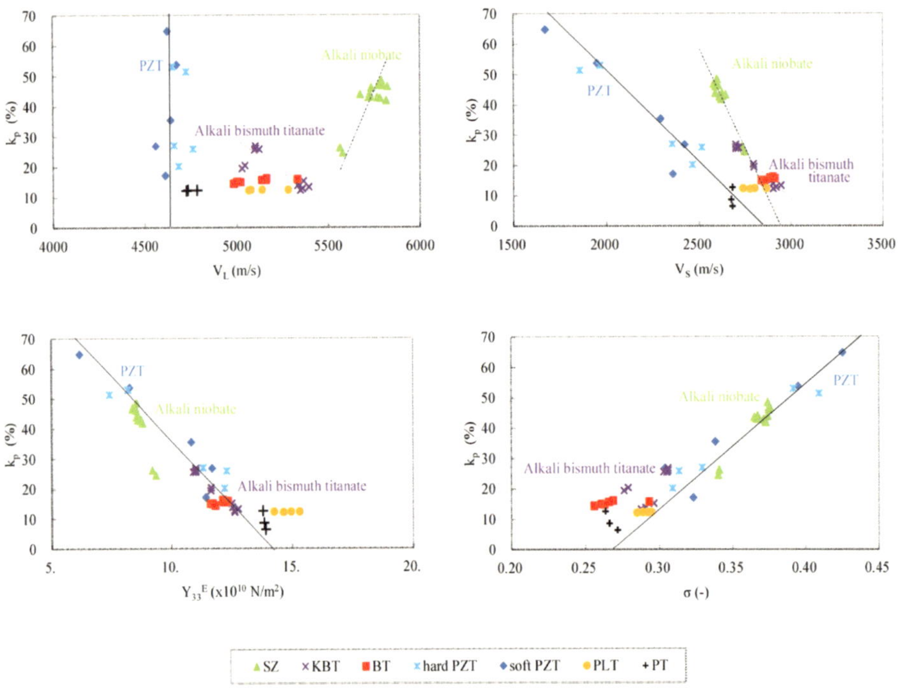

Figure 14. Relationships between longitudinal (V_L) and transverse (V_S) wave velocities, Young's modulus (Y_{33}^E), and Passion ratio (σ) vs planar coupling factors (k_p) in (1-x)(Na,K,Li,Ba) $(Nb_{0.9}Ta_{0.1})O_3$-xSZ (abbreviated to "SZ"), (1-x)NBT-xKBT ("KBT"), 0.79NBT-0.20KBT-0.01BFT ("KBT"), and (1-x)NBT-xBT ("BT") lead-free ceramics compared with $0.05Pb(Sn_{0.5}Sb_{0.5})O_3$-(0.95-x) $PbTiO_3$-x$PbZrO_3$ ceramics with ("hard PZT") and without 0.4 wt% MnO_2 ("soft PZT"), and with $0.90PbTiO_3$-$0.10La_{2/3}TiO_3$ ("PLT") and $0.975PbTiO_3$-$0.025La_{2/3}TiO_3$ ("PT") lead-containing ceramics after full DC poling.

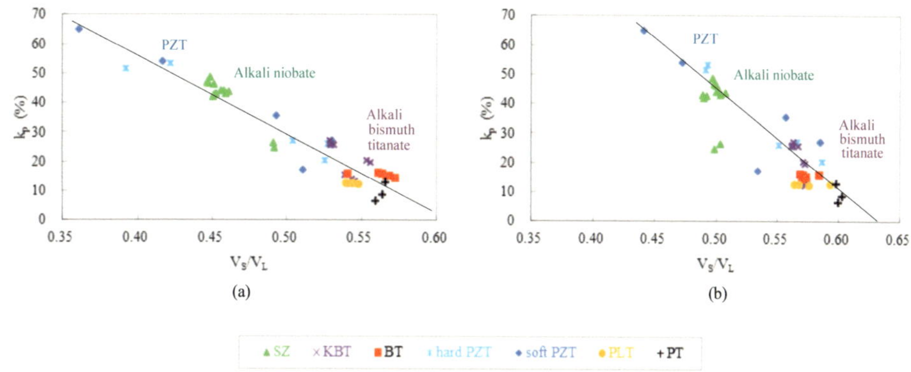

Figure 15. Ratio of longitudinal wave velocity (V_L) to transverse wave velocity (V_S): V_S/V_L vs k_p in lead-containing ceramics [$0.05Pb(Sn_{0.5}Sb_{0.5})O_3$-$(0.95-x)PbTiO_3$-$xPbZrO_3$ with ("hard PZT") and without 0.4 wt% MnO_2 ("soft PZT"), $0.90PbTiO_3$-$0.10La_{2/3}TiO_3$ ("PLT"), and $0.975PbTiO_3$-$0.025La_{2/3}TiO_3$ ("PT")] and in lead-free ceramics [$(1-x)(Na,K,Li,Ba)(Nb_{0.9}Ta_{0.1})O_3$-$x$SZ (abbreviated to "SZ"), (1-x)NBT-xKBT ("KBT"), 0.79NBT-0.20KBT-0.01BFT ("KBT"), and (1-x)NBT-xBT ("BT")] (a) after and (b) before full DC poling.

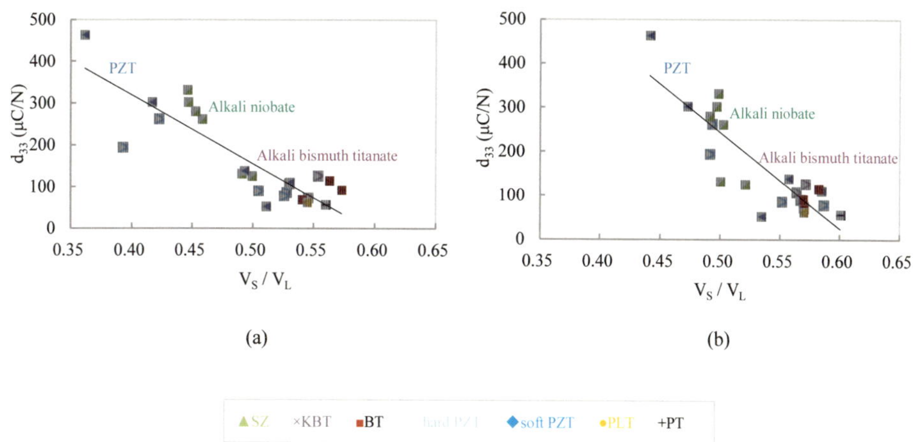

Figure 16. Ratio of longitudinal wave velocity (V_L) to transverse wave velocity (V_S): V_S/V_L vs piezoelectric strain constant d_{33} in lead-containing ceramics ["hard PZT", "soft PZT", "PLT" and "PT"] and in lead-free ceramics ["SZ", "KBT" and "BT"] (a) after and (b) before full DC poling.

Figures 15 and 16 show the ratio of V_L to V_S vs k_p and V_S/V_L vs piezoelectric strain constant d_{33} in lead-containing and lead-free ceramics (a) after and (b) before full DC poling, respectively. There are linear relationships between V_S/V_L vs k_p and d_{33} in spite of DC poling treatment; with decreasing V_S/V_L from 0.58 to 0.36, k_p increased from 4 to 65% (correlation coefficient r=-0.96 after poling and r=-0.89 before poling) and d_{33} also increased from 52 to 463 μC/N (r=-0.87 in both after and before poling). This means that it is possible to estimate the degree of piezoelectricity (the values of k_p and d_{33}) from V_S/V_L, even though the ceramics were without DC poling (as-fired).

Furthermore, we can mention that it is significant to evaluate the both values of k_p and d_{33}, which are not directly corresponding to each other because of $d=k(\varepsilon \cdot s)^{1/2}$ (d: piezoelectric strain constant, k: coupling factor, ε: dielectric constant, s: compliance). Therefore, our developed method for acoustic wave velocity measurement, especially V_S/V_L, is useful for materials research on piezoelectric ceramics including lead-free ceramics with higher coupling factor and higher piezoelectric strain constant d_{33} in the cases of as-fired [Figs. 15(b) and 16(b)] as well as DC fully polarized [Figs. 15(a) and 16(a)] ceramics.

This session was described the road maps in piezoelectric ceramics between longitudinal and transvers wave velocity, Young's modulus and Poisson's ratio to research and develop new piezoelectric ceramic materials, especially lead-free ceramics with high piezoelectricity.

5. Conclusions

Longitudinal and transverse wave velocities of PZT, lead titanate and lead-free ceramics were measured by an ultrasonic precision thickness gauge with high-frequency pulse oscillation to calculate elastic constants such as Young's modulus and Poisson's ratio. The dependences of compositions and manufacturing processes of the ceramics on the elastic constants were clarified in PZT ceramics. Moreover, while the compositions around MPB in lead-free ceramics were investigated by this method, it was found significant relationships between Young's modulus and Poisson's ratio in lead-free ceramics with high piezoelectricity. The DC poling field dependence of the longitudinal and transverse wave velocities of PZT, lead titanate, and lead-free ceramics was also investigated by this method. The effect of the poling field on domain alignment, such as in the cases of full DC poling and domain clamping, could be explained by the relationships between acoustic wave velocities, Young's modulus, and Poisson's ratio vs poling field. The directions of the research and development of piezoelectric ceramics including lead-free ceramics with high coupling factors could be proposed on the basis of the findings of this study.

Acknowledgments

This work was partially supported by a Grant-in-Aid for Scientific Research C (No. 21560340) and a Grant of Strategic Research Foundation Grant-aided Project for Private Universities 2010-2014 (No. S1001032) from the Ministry of Education, Culture, Sports, Science and Technology, Japan (MEXT).

Author details

Toshio Ogawa

Department of Electrical and Electronic Engineering, Shizuoka Institute of Science and Technology, Toyosawa, Fukuroi, Shizuoka, Japan

References

[1] Ogawa T, Yamada A, Chung YK, Chun DI. Effect of Domain Structures on Electrical Properties in Tetragonal PZT Ceramics. J. Korean Phys. Soc. 1998;32: S724-S726.

[2] Ogawa T, Nakamura K. Poling Field Dependence of Ferroelectric Properties and Crystal Orientation in Rhombohedral Lead Zirconate Titanate Ceramics. Jpn. J. Appl. Phys. 1998;37: 5241-5245.

[3] Ogawa T, Nakamura K. Effect of Domain Switching and Rotation on Dielectric and Piezoelectric Properties in Lead Zirconate Titanate Ceramics. Jpn. J. Appl. Phys. 1999;38: 5465-5469.

[4] Ogawa T. Domain Switching and Rotation in Lead Zirconate Titanate Ceramics by Poling Fields. Ferroelectrics 2000;240: 75-82.

[5] Ogawa T. Domain Structure of Ferroelectric Ceramics. Ceram. Int. 2000;25: 383-390.

[6] Ogawa T. Poling Field Dependence of Crystal Orientation and Ferroelectric Properties in Lead Titanate Ceramics. Jpn. J. Appl. Phys. 2000;39: 5538-5541.

[7] Ogawa T. Poling Field Dependence of Ferroelectric Properties in Barium Titanate Ceramics. Jpn. J. Appl. Phys. 2001;40: 5630-5633.

[8] Ogawa T. Poling Field Dependence of Ferroelectric Properties in Piezoelectric Ceramics and Single Crystal. Ferroelectrics 2002;273: 371-376.

[9] Kato R, Ogawa T. Chemical Composition Dependence of Giant Piezoelectricity on k_{31} Mode in $Pb(Mg_{1/3}Nb_{2/3})O_3$–$PbTiO_3$ Single Crystals. Jpn. J. Appl. Phys. 2006; 45: 7418-7421.

[10] Ogawa T, Yamauchi Y, Numamoto Y, Matsushita M, Tachi Y. Giant Electromechanical Coupling Factor of k_{31} Mode and Piezoelectric d_{31} Constant in $Pb[(Zn_{1/3}Nb_{2/3})_{0.91}Ti_{0.09}]O_3$ Piezoelectric Single Crystal. Jpn. J. Appl. Phys. 2002;41: L55-L57.

[11] Ogawa T. Giant k_{31} Relaxor Single-Crystal Plate and Their Applications. In: Lallart M. (ed.) Ferroelectrics-Applications. Rijeka: InTech; 2011. P.3-34.

[12] Ogawa T, Numamoto Y. Origin of Giant Electromechanical Coupling Factor of k_{31} Mode and Piezoelectric d_{31} Constant in $Pb[(Zn_{1/3}Nb_{2/3})_{0.91}Ti_{0.09}]O_3$ Single Crystal. Jpn. J. Appl. Phys. 2002;41: 7108-7112.

[13] Technical catalogues 2004 for piezoelectric materials of TDK Corporation, Murata Manufacturing Co., Ltd., Fuji Ceramics Corporation, NEC Tokin Corporation, etc.

[14] Kadota M, Saito Y, Yoshida T, Ozeki H, Yamaguchi T, Hasegawa M, Sato H. EMAS-6100, Standard of Electronic Materials Manufacturers Association of Japan. Tokyo: Piezoelectric Ceramics Technical Committee; 1993 [in Japanese].

[15] Bechmann R. IRE Standards on Piezoelectric Crystals. Proc. IRE, 1958.

[16] Shibayama K. Danseiha Soshi Gijyutsu Handbook (Technical Handbook of Acoustic Wave Devices). Tokyo: Ohmsha; 1991 [in Japanese].

[17] Mason WP. Physical Acoustics I, Part A. New York: Academic Press; 1964.

[18] Takagi K. Cyouonpa Binran (Ultrasonic Handbook). Tokyo: Maruzen; 1999 [in Japanese].

[19] Negishi K, Takagi K. Cyouonpa Gijutsu (Ultrasonic Technology). Tokyo: Univ. of Tokyo Press; 1984 [in Japanese].

[20] Furukawa M, Tsukada T, Tanaka D, Sakamoto N. Alkaline Niobate-based Lead-Free Piezoelectric Ceramics. Proc. 24th Int. Japan-Korea Semin. Ceramics, 2007.

[21] Ogawa T, Furukawa M, Tsukada T. Poling Field Dependence of Piezoelectric Properties and Hysteresis Loops of Polarization versus Electric Field in Alkali Niobate Ceramics. Jpn. J. Appl. Phys. 2009;48, 709KD07-1-5.

[22] Ogawa T, Nishina T, Furukawa M, Tsukada T. Poling Field Dependence of Ferroelectric Properties in Alkali Bismuth Titanate Lead-Free Ceramics. Jpn. J. Appl. Phys. 2010; 49: 09MD07-1-4.

[23] Ogawa T. Highly Functional and High-Performance Piezoelectric Ceramics. Ceramic Bulletin 1991;70: 1042-1049.

[24] Zhao W, Zhou H, Yan Y, Liu D. Morphotropic Phase Boundary Study of the BNT-BKT Lead-Free Piezoelectric Ceramics. Key Eng. Mater. 2008;368-372: 1908-1910.

[25] Yang Z, Liu B, Wei L, Hou Y. Structure and Electrical Properties of $(1\text{-}x)Bi_{0.5}Na_{0.5}TiO_3$-$xBi_{0.5}K_{0.5}TiO_3$ Ceramics near Morphotropic Phase Boundary. Mater. Res. Bull. 2008;43: 81-89.

[26] Dai Y, Pan J,Zhang X. Composition Range of Morphotropic Phase Boundary and Electrical Properties of NBT-BT System. Key Eng. Mater. 2007; 336-338: 206-209.

[27] Saito Y, Takao H, Tani T, Nonoyama T, Takatori K, Homma T, Nagaya T, Nakamura M. Lead-Free Piezoceramics. Nature 2004;432: 84-87.

http://dx.doi.org/10.5772/53117

[28] Jaffe B, Cook WR, and Jaffe H. Piezoelectric Ceramics. New York: Academic Press; 1971.

Chapter 3

The Application of Piezoelectric Materials in Machining Processes

Saeed Assarzadeh and Majid Ghoreishi

Additional information is available at the end of the chapter

http://dx.doi.org/10.5772/54579

1. Introduction

In 1880, Jacques and Pierre Curie discovered that pressure generates electrical charges in a number of crystals such as Quartz and Tourmaline, calling this phenomenon the "piezoelectric effect". Later, they noticed that electrical fields can also deform piezoelectric materials, showing an inverse effect comparing to their first observation. Nowadays, the effect has gained considerable practical attractions in miscellaneous industrial applications including machining processes as a broad range within manufacturing methods in our competitively global technologies.

This chapter focuses mainly on the usages of piezoelectric materials in two differently common practiced machining operations adopted by both manufacturers and research scholars.

The first study pays attention to the role of piezoelectric transducers as the core system in producing ultrasonic waves in ultrasonic machines. Ultrasonic machining (USM) process, as one of the popular nontraditional machining processes, is capable of chip removal from every brittle material, whether conductive or nonconductive, susceptible to failure under mechanical loads in conventional machining processes of whatsoever hardness, such as ceramics, glass, porcelains, etc. The viability and effectiveness of piezoelectric transformers with high rate of electro-mechanical conversion compared to their old magnetostrictive counterparts are described and analyzed. The USM process capabilities and applications are also succinctly introduced.

The second is the application of quartz as a piezoelectric material in dynamometers used to measure forces and torques during conventional machining processes, like turning, milling, drilling, and so on. The basic principles and features of how a piezoelectric-based dynamometer works are discussed along with the need to measure forces and torques through dynamometry.

2. Ultrasonic machining (USM) of materials

Ultrasonic machining is an economically viable operation by which a hole or a cavity can be pierced in hard and brittle materials, whether electric conductive or not, using an axially oscillating tool. The tool oscillates with small amplitude of 10-15 μm at high frequencies of 18-40 KHz to avoid unnecessary noise and being above the upper frequency limit of the human ear, justifying the tem "ultrasonic" [1, 2].

During tool oscillation, abrasive slurry (B4C and SiC) is continuously fed into the working gap between the oscillating tool and the stationary WP. The abrasive particles are, therefore, hammered by the tool into the WP surface, and consequently abrading the WP into a conjugate image of the tool form. Moreover, the tool imposes a static pressure ranging from 1N to some kilograms depending on the size of the tool tip, see Fig. 1. This static pressure is necessary to sustain the tool feed during machining. Owing to the fact that the tool oscillates and moves axially, USM is not limited to the production of circular holes. The tool can be made to the shape required, and hence extremely complicated shapes can be produced in hard materials. Beside machining domain, US techniques are applied in nondestructive testing (NDT), welding, and surface cleaning, as well as diagnostic and medical applications.

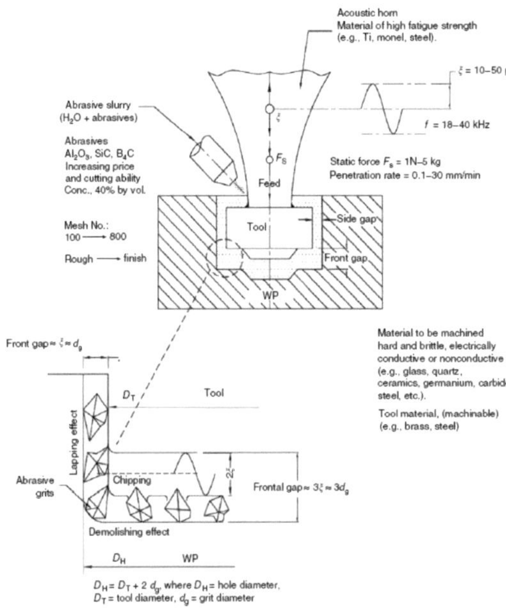

Figure 1. Characteristics of the USM process [1]

2.1. Elements of process

The USM equipment shown in Fig. 2 has a table capable of orthogonal displacement in X and Y directions, and a tool spindle and carrying the oscillating system moving in direction Z

http://dx.doi.org/10.5772/54579

perpendicular to the X-Y plane. The machine is equipped with a HF generator of a rating power of 600 W, and a two-channel recording facility to monitor important machining variables (tool displacement Z and oscillation amplitude ξ). A centrifugal pump is used to supplement the abrasive slurry into the working zone. Fig. 3 shows schematically the main elements of the equipment, which consist of the oscillating system, the tool feeding mechanism, and the slurry system.

Figure 2. USM equipment [1]

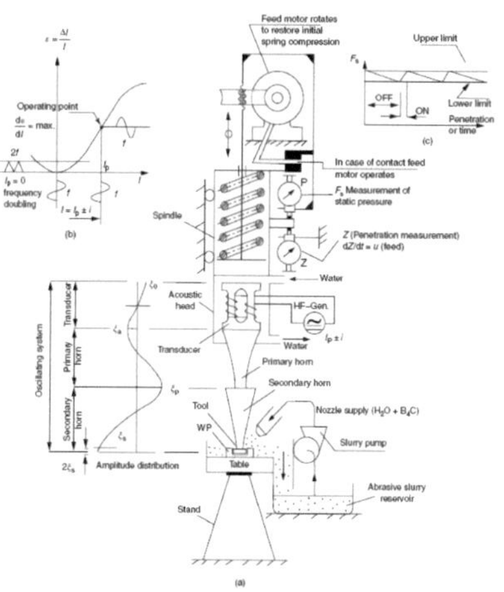

Figure 3. The schematic of complete vertical USM equipment [1]

2.1.1. Oscillating system and magnetostriction effect

The core element of each US machine, the oscillating system, includes a transducer in the acoustic head, a primary horn, and a secondary acoustic horn (see Fig. 4).

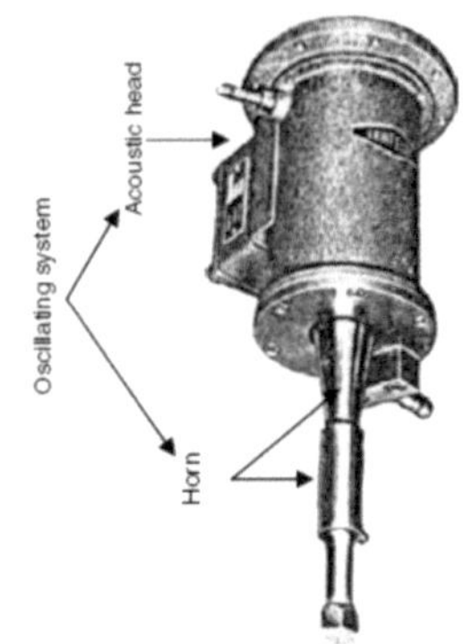

Figure 4. The oscillating system of USM equipment [1]

2.1.2. Acoustic transducer

This transforms electrical energy to mechanical energy in the form of oscillations. Magnetostrictive transducers are generally employed in USM, but piezoelectric ones may also be used.

The magnetostriction effect was first discovered by Joule in 1874. According to this effect, in the presence of an applied magnetic field, ferromagnetic metals and alloys change in length. The deformation can be positive or negative, depending on the ferromagnetic material. An electric signal of US-frequency f_r is fed in to a coil that is wrapped around a stack made of magnetostrictive material (iron-nickel alloy). This stack is made of laminates to minimize eddy current and hysteresis losses; moreover, it must be cooled to dissipate the generated heat (Fig. 3a). The alternating magnetic field produced by the HF-ac generator causes the stack to expand and contract at the same frequency.

To achieve the maximum magnetostriction effect, the HF-ac current i must be superimposed on an appropriate dc premagnetizing current I_p that must be exactly adjusted to attain an optimum or working point. This point corresponds to the inflection point ($d^2\varepsilon/dl^2=0$) of the magnetostriction curve, (Fig 3b). Without the application of premagnetizing direct current I_p, it is evident that the magnetostriction effect occurs in the same direction for a given ferromagnetic material irrespective of the field polarity, and hence the deformation will vary at twice the frequency $2f_r$ of the oscillating current providing the magnetic field (Fig. 3b). Therefore, the premagnetizing direct current I_p has the following functions:

- When precisely adjusted, it provides the maximum magnetostriction effect (maximum oscillating amplitude)
- It prevents the frequency doubling phenomenon

If the frequency of the ac signal, and hence that of the magnetic field, is tuned to be the same as the natural frequency of the transducer (and the whole oscillating system), so that it will be at mechanical resonance, then the resulting oscillation amplitude becomes quite large and the exciting power attains its maximum value.

http://dx.doi.org/10.5772/54579

2.1.3. *Transducer length*

The resonance condition is realized if the transducer length is *l*, which is equal to half of the wave length, λ (or positive integer number *n* of it).

Therefore,

$$l = \frac{n}{2}\lambda = \frac{\lambda}{2}, \ if \ n = 1$$

and

$$\lambda = \frac{c}{f_r} = \frac{1}{f_r}\sqrt{\frac{E}{\rho}}$$

where

c = acoustic speed in magnetostrictive materials (m/s)

f_r = resonant frequency (1/s)

E, ρ = Young's modulus (MPa) and density (kg/m^3) of the magnetostrictive material

Hence,

$$l = \frac{c}{2f_r} = \frac{1}{2f_r}\sqrt{\frac{E}{\rho}}$$

2.1.4. *Piezoelectric transducers*

A main drawback of magnetostriction transducers is the high power loss ($\eta = 55\%$). The power loss is converted into heat, which necessitates the cooling of the transducer. In contrast, piezoelectric transducers are more efficient ($\eta = 90\%$), even at higher frequencies (f = 25-40 KHz). Piezoelectric transducers utilize crystals like quartz and lead titanate-zirconate that undergoe dimensional changes proportional to the voltage applied. Similar to magnetostrictors, the length of crystal should be equal to half the wavelength of the sound in the crystal to produce resonant condition. At a frequency of 40 KHz, the resonant length *l* of the quartz crystal ($E = 5.2 \times 10^4$ MPa, $\varrho = 2.6 \times 10^3$ kg/m^3) is equal to 57 mm. Sometimes a polycrystalline ceramic like barium titanate is used.

Piezoelectric transformers were first introduced into modern ultrasonic machines in the late 1960s. In 1970, Tyrrell [3] has described such a system. In essence, piezoelectric transducers are composed of small particles bound together by sintering; undergoing polarization by heating above the Curie point and placing it in an electric field such that orientation is preserved on cooling. A disc of the piezoelectric material which has a very high electromechanical conversion rating is sandwiched between two thick metal plates to form the ultrasonic horn. When a current of fixed frequency is fed to the horn the whole system is found to vibrate at some resonant frequency along the longitudinal axis; acoustically the motion is equal to one half a wavelength.

2.2. The advantages and disadvantages of USM process [4, 5]

Advantages

Some of the special priorities can be mentioned as follow:

- Intricate and complex shapes and cavities in both electric and nonelectric materials can be readily machined ultrasonically
- As the tool exhibits no rotational movement, the process is not limited to produce circular holes
- High dimensional accuracy and surface quality
- Especially, in the sector of electrically nonconductive materials, the USM process is not in competition with other nontraditional machining processes regarding accuracy and removal rates
- Since there is no temperature rise of the WP, no changes in physical properties or micro-structure whatsoever can be expected

However, the USM process has some disadvantages listing below.

Disadvantages:

- The USM is not capable of machining holes and cavities with a lateral extension of more than 25-30 mm with a limited depth of cut
- The tool suffers excessive frontal and side wear when machining conductive materials such as steels and carbides. The side wear destroys the accuracy of holes and cavities, leading to a considerable conicity error.
- Every job needs a special high-cost tool, which adds to the machining cost
- High rate of power consumption
- In case of blind holes, the designer should not allow sharp corners, because these cannot be produced by the USM.

2.3. The applications of USM

It should be understood that the USM is generally applied to machining shallow cavities and forms in hard and brittle materials having a surface area not more than 10 cm^2. Some typical applications of USM are as follow:

- Manufacturing forming dies in hardened steel and sintered carbides
- Manufacturing wire drawing dies, cutting nozzles for jet machining applications in sapphire, and sintered carbides
- Slicing hard brittle materials such as glass, ceramics, and carbides
- Coining and engraving applications

- Boring, sinking, blanking,, and trepanning
- Thread cutting in ceramics by rotating the tool or the WP

Figure 5 illustrates some of the products produced ultrasonically.

a. Engraving a medal made of agate

b. Piercing and blanking of glass

c. Producing a fragile graphite electrode for EDM

d. Sinking a shearing die in hardened steel or WC

e. Production of outside contour and holes of master cutters made of zirconium oxide (ZrO_2) of a textile machine

f. Drilling fine holes Φ = 0.4 mm in glass

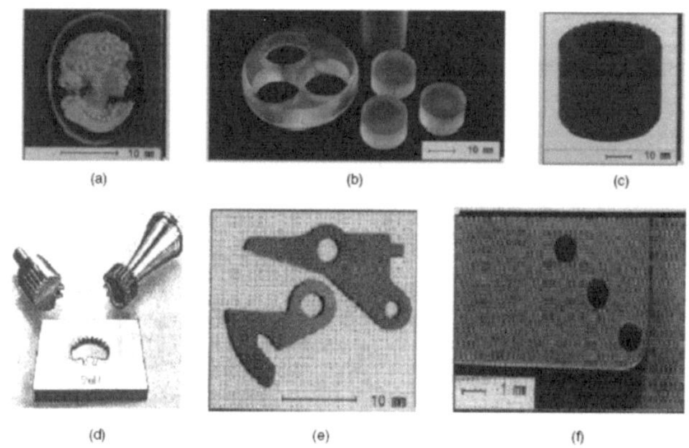

Figure 5. Some typical products by USM [1]

3. Dynamometry in conventional machining processes

During machining, the cutting tool exerts a force on the WP as it removes the machining allowance in the form of chips. Empirical values for estimating the cutting forces are no longer sufficient to reliably establish the optimum machining conditions. Depth of cut, feed rate, cutting speed, WP materials, tool material and geometry, and cutting fluid are just a few of the machining parameters governing the amplitude and direction of the cutting force.

The optimization of a machining process necessitates accurate measurement of the cutting force by a special device called a machine tool dynamometer, capable of measuring the components of the cutting force in a given coordinate system. It is a useful and powerful tool employed in a variety of applications in engineering research and manufacturing. A few examples of these applications are [1]:

- Investigating into the machinability of materials
- Comparing similar materials from different sources
- Comparing and selecting cutting tools
- Determining optimum machining conditions
- Analyzing causes of tool failure
- Investigating the most suitable cutting fluids
- Determining the conditions that yield the best surface quality
- Establishing the effect og fluctuating cutting forces on tool wear and tool life

The machine tool dynamometer is not standard equipment or a device that can be used on every machine. Rather, it is equipment especially designed to fulfill some desired requirements that adapt a specific machine type operating at a specific range of machining conditions.

3.1. Piezoelectric (Quartz) dynamometers

Of the numerous piezoelectric materials, quartz is by far the most suitable one for force measurement, because it is stable material with constant properties. In its crystalline form, quartz is anisotropic, in that its material properties are not identical in all directions. Depending on the position in which they are cut out of the crystal, disks are obtained that are:

1. Sensitive only to pressure (longitudinal effect), Fig. 6a, which measure the main force component F_z (brown)
2. Sensitive only to shear in one particular direction (shear effect), Fig. 6b, which measures components F_x (blue) and F_y (green), perpendicular to F_z, as well as the torque M_z (red). Figure 6c illustrates the generalized multi-components with reference to a Cartesian coordinate system.

The piezoelectric force measuring principle differs fundamentally from their old traditional counterparts, the strain and displacement based dynamometers, in that it is an active system. When a force acts on a quartz element, a proportional electric charge appears on the loaded surfaces, meaning that it is not necessary to measure the actual deformation.

In piezoelectric dynamometers, the deflection is not more than a few micrometers at full load, whereas with conventional systems, several tenths of a millimeter may be needed. Thus, piezoelectric dynamometers are very stiff systems and their resonant frequency is high, so that even rapid events can be measured satisfactorily. Moreover, the individual components of the cutting force can be measured directly, eliminating any interference between measuring channels. Quartz dynamometers require no zero adjustment or balancing of the bridge circuit. It is just a matter of pressing a button, being ready for duty. The outstanding features of quartz dynamometers are [1, 6]:

- High rigidity, hence high resonant frequency

- Minimal deflections (few micrometers at full load)
- Wide measuring range
- Linear characteristics, free of hysteresis
- Lowest cross talk (typically under 1%)
- Simple in operation and without need for bridge balancing
- Compact design
- Unlimited life expectancy

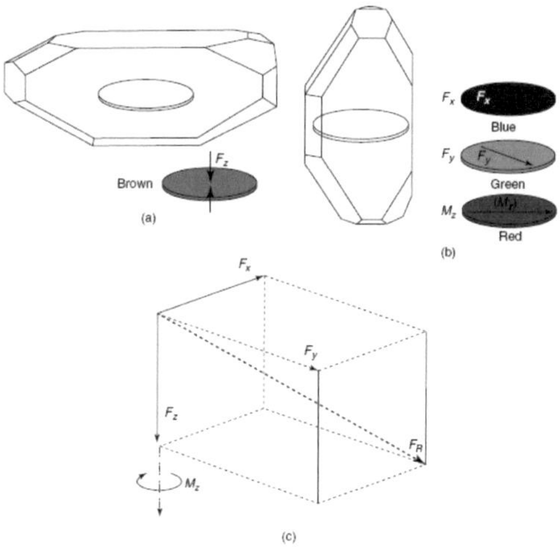

Figure 6. Disks of quartz crystals. (a) Pressure-sensitive, (b) shear-sensitive, (c) multi-components in reference to a Cartesian coordinate system [6]

3.2. Typical piezoelectric dynamometers

Piezoelectric dynamometers are efficiently used on the majority of machine tools. Three application examples are described below with their corresponding setup.

1. *Two-component piezoelectric drilling dynamometers.*Figure 7 illustrates a two-component drilling dynamometer in which shear-sensitive discs are organized in a circle with their shear-sensitive axes oriented to respond to the torque M_z (red), whereas pressure-sensitive disks are arranged and oriented to measure the trust load F_z (brown). A high

preload is necessary because the shear forces must be transmitted by friction to measure the torque.

The two-component dynamometer, shown in Fig. 7, is suited for operations including drilling, thread cutting, countersinking, reaming, and so on. Torques and forces acting when machining holes from less than 1 mm to over 20 mm diameter can be measured satisfactorily by this dynamometer. A record of M_z and F_z is illustrated in Fig. 7, from which it is clearly seen that F_z rises steeply at the beginning (entry of tool chisel), followed by the gradual rise of the M_z component, as the latter is more affected by the force acting on the two drill lips.

2. *Three-component piezoelectric turning dynamometer.* This model includes several shear-sensitive quartz, with their shear-sensitive axes oriented to measure F_x (blue ring) and F_y (green ring), respectively. Their shear sensitive-axes are inclined to each other at an angle of 90^0, and both are contained in a housing to form a two-component force measurement element for F_x and F_y (Figure 8). Pressure-sensitive quartz disks are contained in a single housing to form a single-component force-measuring element for F_z (brown ring). Another alternative is illustrated in the construction shown in Figure 8, where three separate elements for measuring F_x, F_y, and F_z are sandwiched under high preload between a base plate and a top plate. The dynamometer is mounted on the lathe slide in place of a cross-slide. A record of the three components is shown also in the same figure, from which it is clear that F_x=F_y, meaning that the cut is preferred at an approach angle χ=45^0.

3. *Three-component piezoelectric milling or grinding dynamometer.* Whole quartz rings may be employed. Two-shear-sensitive quartz pairs, for F_x (blue) and F_y (green), and a pressure-sensitive pair for F_z (brown), can be assembled in a common housing to form a three-component force-measuring element (Figure 9). The pressure-sensitive quartz are arranged in the middle so that they lie in the neutral axis under bending. During milling and grinding, the application point of the force varies a great deal. Consequently, dynamometers having four piece three-component force-measuring elements are employed. All the x, y, and z channels respectively are paralleled electrically. This makes the measurement independent of the momentary force application point. For bigger work, two dynamometers paralleled electrically and mechanically may be employed together. This system measures correctly independent of the point of force application. A typical output of the three-component milling dynamometer is shown in Fig. 9. The milling process has been performed under the following conditions:

- Status: Up milling
- Cutter diameter = 63 mm, helix β = 30^0, n = 90 rpm, Z = 12 teeth
- Feed u = 53 mm/min
- Depth of cut t = 3.5 mm

The severe periodic fluctuation in the measured forces is attributed to an eccentric motion of the cutter shaft. Superimposed are vibrations due to gearing of the machine. It is perfectly clear from the record that the setup shown is far from ideal. The force measure, therefore, sheds light on the machine tool behavior as well, and not just on the actual cutting operation.

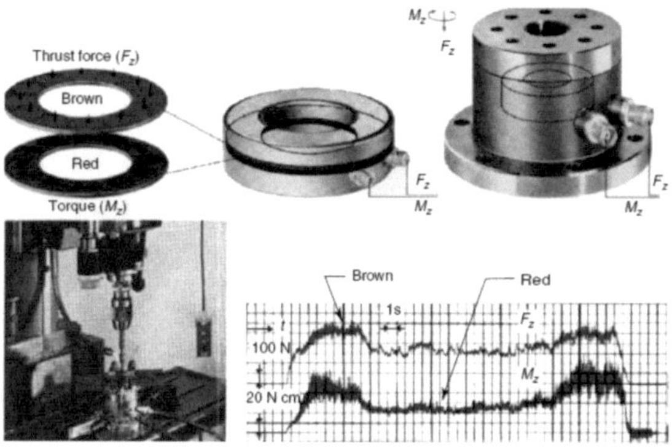

Figure 7. Two-component piezo drilling dynamometer [6]

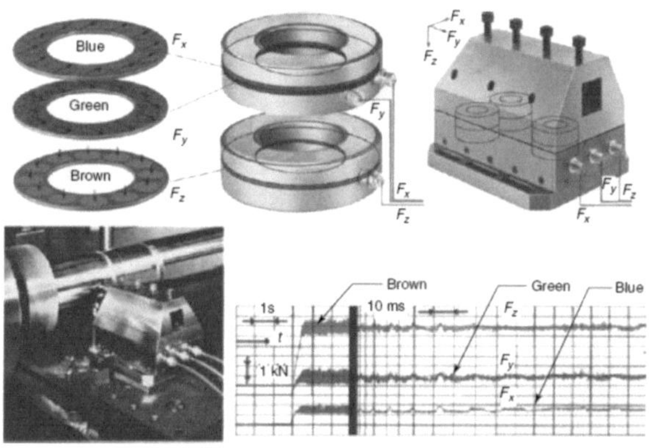

Figure 8. Three-component piezo turning dynamometer [6]

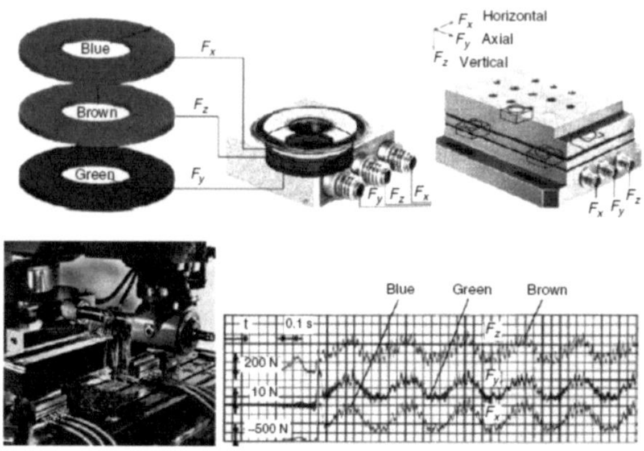

Figure 9. Three-component piezo milling dynamometer [6]

4. Conclusion

In this chapter, attempt has been made to cover the basic details of utilizing piezoelectric materials in two different fields of machining processes. The first focused on the effectiveness of a piezoelectric transducer in producing ultrasonic waves as a cutting tool in ultrasonic machining processes. Compared to their old magnetostriction counterparts, the piezo ones demonstrate higher electro-mechanical efficiency. Besides, they are more compact as well as being simpler in design and operation. The second emphasizes the unique suitability of piezoelectric dynamometers in measuring various components of forces and torques generated during different kinds of conventional machining processes. The importance of measuring forces and torques in an on-line manner during every traditional machining operation for the purposes of modeling and optimization makes the use of a precise piezoelectric dynamometer an inseparable part in manufacturing industries as well as academic domains.

Author details

Saeed Assarzadeh* and Majid Ghoreishi

*Address all correspondence to: saeed_assarzadeh@yahoo.com

Department of Mechanical Engineering, K. N. Toosi University of Technology, Tehran, Iran

References

[1] Youssef Helmi A., El-Hofy, H., Machining technology: Machine tools and operations, CRC Press, (2008). 978-1-42004-339-6

[2] Mcgeough, J. A. Advanced methods of machining, Chapman and Hall, London and New York, (1988). 0-41231-970-5

[3] Tyrrell, W. R. Rotary ultrasonic machining, Soc. Manufacturing Eng., USA, Paper (1970). (MR70-516), 1-10.

[4] El-Hofy, H. Advanced machining processes: Nontraditional and hybrid machining processes, McGraw Hill Book Com., (2005). 0-07145-334-2

[5] Pandey, P. C, & Shan, H. S. Modern machining processes, Tata McGraw Hill Book Com., (2004). 0-07096-553-6

[6] http://www.Kistler.com

Chapter 4

Application of the Piezoelectricity in an Active and Passive Health Monitoring System

Sébastien Grondel and Christophe Delebarre

Additional information is available at the end of the chapter

http://dx.doi.org/10.5772/54581

1. Introduction

Fibber reinforced composites are nowadays used extensively in aircraft structures because of their properties such as low weight, high stiffness, high strength and fatigue resistance. Nevertheless, they are not exempt from drawbacks, since they are very sensitive to manufacturing processes and service conditions. In particular, their high weakness to low and high velocity impacts has brought new problems for maintenance. These events that are prime sources of delamination and fibber cracking in composite structures are produced either by hazardous conditions (e.g. bird strikes, impacts with foreign objects, etc.) or human errors (tool drops, ground collisions, etc.). In this context, the development of a continuous health monitoring in parallel to the traditional maintenance is a safety issue [1].

Two main strategies are possible to monitor the structural health of composite structures. The first one is the detection of the damaging event continuously, i.e. the detection of the Acoustic Emission (AE) energy that is generally released when the material is bent or cracks due to an external load (pressure, impact, temperature, etc…). This strategy needs permanent monitoring, in flight and on the ground as well [2-4]. The second strategy consists in detecting the damage itself by periodically checking the structural health. Damage detection is then made with help of comparison of the initial state to the actual one. In this situation, the health monitoring system can be either local or global. For the local inspection, the sensor must be in the damaged region, registering permanent strains due to the damage [5-6]. For the global inspection, stimulations are produced in view to induce a structural response, analysed by the sensors. These stimulations can excite the full structure for modal [7] or static analysis [1], or only a small region for the acousto-ultrasonic technique [8-9].

In order to improve the health monitoring system, some scientists have proposed to combine the previously discussed techniques together. Hence, [10] was one of the first to use the same piezoelectric transducers in order to perform a passive diagnosis (PSD) and an active sensing diagnosis (ASD). The PSD utilized the sensor measurement to determine the impact force and predict the impact location, whereas the ASD generated diagnostic signals from the actuators to estimate the size of the impact damage. Similarly, [11] demonstrated the possibility to develop an health monitoring system based first on the excitation and reception of guided waves along the structure by using thin piezoelectric transducers (active mode) and secondly on a continuous monitoring taking the same transducers used as AE sensors (passive mode). Their goal was to monitor disbond growth and damaging impact in a composite wingbox structure. To increase the system sensitivity, [12] tried to combine high frequency propagating elastic waves with low-frequency vibrations. This technique also called vibro-ultrasonic technique allows, by applying an additional low frequency, to move the damage, i.e. to open and close crack or delamination. As a result, the high frequency ultrasonic waves are modulated due to varying size of the damage, the intensity of the modulation being proportional to the severity of the damage. [13] proposed the coupling of an electromagnetic sensors network and ceramic piezoelectric sensors. The electromagnetic method is particularly sensitive to local burning, fibber cracking and liquid ingress, whereas the acousto-ultrasonic method is more sensitive to mechanical damage such as delaminations. [14] utilized impedance (local inspection) and guided wave (global inspection) based damage detection techniques simultaneously from surface-mounted piezoelectric transducers to enhance the performance and reliability of damage diagnosis especially under varying temperature conditions. Finally, all these studies demonstrated that the use of complementary techniques tend to extend the detection capability while reducing false alarms.

Despite the extensive literature on the subject, commercial applications of health monitoring systems for damage detection were applied principally to one-dimensional structures such as pipes, and rails, and simple structures like plates. There are two major reasons for this. Firstly, the use of acousto-ultrasonic or acoustic emission techniques in complex structures such as airframes is very complicated due to multiple reflections and mode conversions at features such as ribs and stiffeners which generate signals that are very difficult to interpret. Secondly, many of the proposed methods require a large number of transducers for the monitoring of large structures; this is often not possible or acceptable. Therefore the principal aim of this chapter is to demonstrate the feasibility of using a passive and active system based on few thin piezoelectric transducers to monitor large and complex structures submitted to a series of damaging impacts. The challenge is to detect, analyze and locate damaging impacts with a minimum number of transducers.

2. Experimental procedure

2.1. Structure description

The tests were conducted on a composite wing-box structure (see Figure 1) specially manufactured in order to be representative of an aircraft wing. Hence, wingbox skins were bolted onto a metallic substructure consisting of three metallic spars. The skins used were rectangular panels (1800*760 mm^2) with a thickness varying from 6 mm to 4 mm and they were made of 913C-HTA composite material. The lay-up and geometry and the material data are given in Appendix A and B, respectively.

Otherwise, as shown in Figure 1(a), the skins had four bonded stringer which were formed around foam cores. Stringer height was 30 mm; width 20 mm and the feet of the stringers were 10 mm wide. The stringers were bonded onto the skins using REDUX 319A structural adhesive.

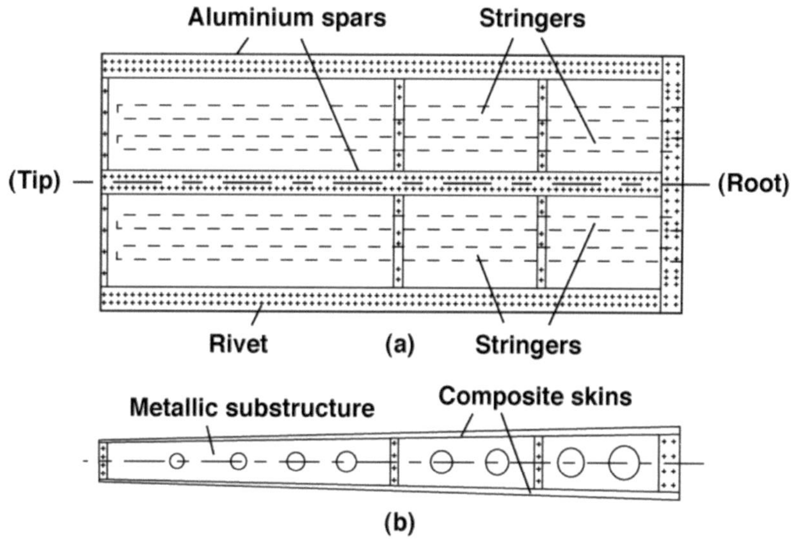

Figure 1. Schematic description of the wing-box structure: (a) Front view; (b) Side view

2.2. Presentation of impact tests

To perform the impacts on the surface of the structure, a mobile tower serving as a guide for the impinger was instrumented with a force sensor. This device thus allowed recording the impact force. In addition, it was possible to apply different energy impacts depending on the height of the fall and the mass of the impinger.

Series of impacts were applied with increasing energy level at three different locations of the structure (see sections 2.3 and 3.4) and after each impact the skin was examined using a manually C-scan ultrasonic system. At the impact location 1, a series of impacts with energy level equal to 6J, 10J, 20J, 30J and 40J was necessary before obtaining a damage of the structure. Using this information, only, two impacts with energy level equal to 6J and 40J were applied at the location 2. Finally, the structure was subjected to successive impacts with energy level equal to 6J, 35J and 40J at the third location.

Figure 2(a) shows the impinger machine while Figures 2(b) and 2(c) illustrate the responses of the force sensor to impacts with energy levels equal to 30J and 40J at location 1. We can notice that the shapes of the force signals are different. Indeed, high frequencies are visible between

2 and 3.5ms on the signal force measured at 40J. The use of the C-Scan described above confirmed that the high frequencies were related to the occurrence of the damage. We will see that this feature can also be used in the analysis of AE signals (see sections 2.5 and 3.2).

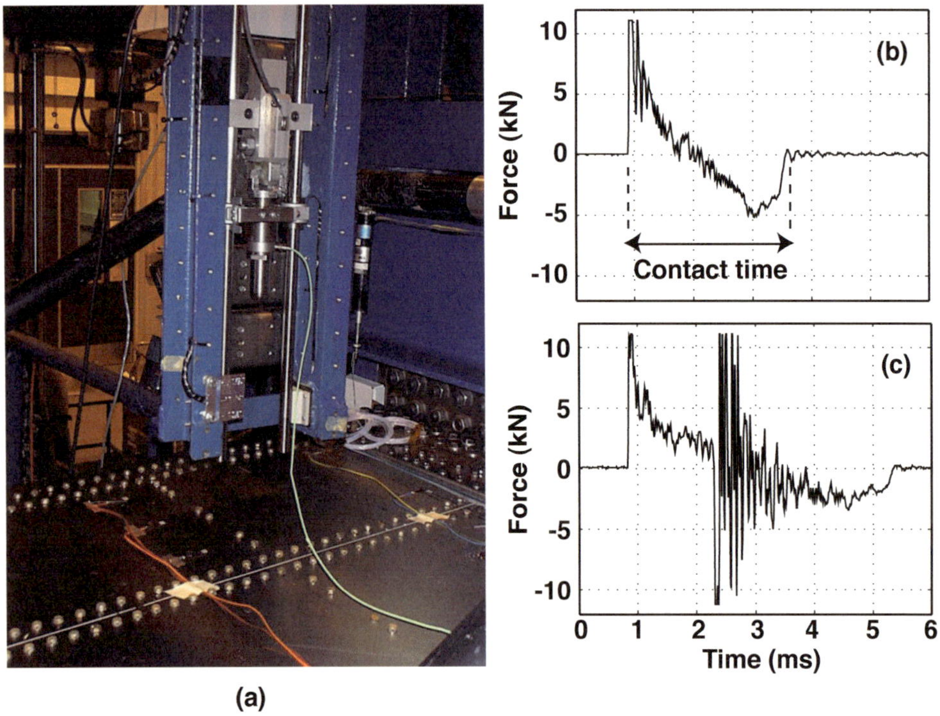

Figure 2. Impinger photography (a), force signal for impact site 1 with energy levels equal to 30J (b) and 40J (c), respectively

2.3. Integration of the health monitoring system

Before the integration of the health monitoring system to the structure, the first task to achieve was to choose an appropriate transducer, i.e. a transducer which could be used complementary to measure either the stress waves generated by damaging impacts (passive mode) or to produce stimulations at discrete time intervals for active health monitoring of the structure (active mode). Since the use of traditional angle probe is not totally fitted because they cannot be permanently fixed on the structure, it was decided to work with low thickness ceramics made of piezoelectric material P1-60 (a standard 'Quartz and Silice' piezoelectric ceramic) and polarized along the thickness. The scaling down of the transducer, particularly the thickness, has the additional advantage of being more adapted to the development of self-monitoring material.

To allow a better directivity of the stimulation in active mode, a rectangular shape for the transducer was privileged. We will see in Section 3.4 that this choice does not affect the results of damage location in passive mode. Moreover, a general rule [15] of ultrasound emission is to excite the emitting element at its natural resonances rather than at any frequency because this method enables a very efficient conversion from electric to mechanic energy. It also means that special care must be taken to choose the thickness, width and length of the piezoelectric elements. Nevertheless, the height of the piezoelectric elements being chosen small (order of 1 mm), working at the thickness resonance (around 1.8 MHz) is not suitable since it does not correspond to the frequency under study. This also motivated our interest to the transverse resonance. Hence, for application of the stimulation, the dimensions of the transducer have been chosen equal to 1*6*30 mm^3 (see Figure 3(a)), the width resonance corresponding approximately to the frequency of interest according to the properties of the P1-60.

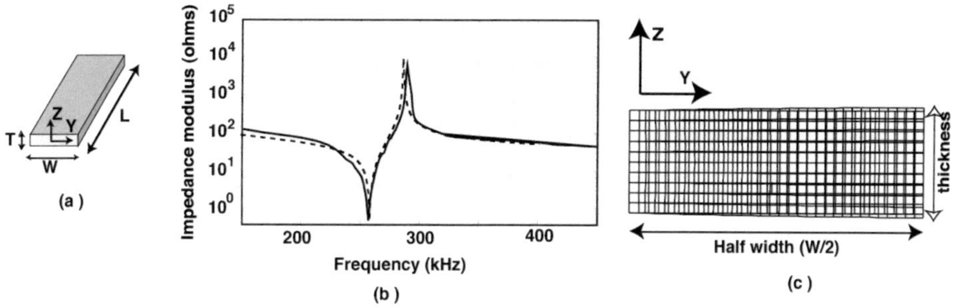

Figure 3. a) Piezoelectric transducer bar, and (b) experimental and computed electrical impedance modulus of the transducer in vacuum, as a function of frequency. Experimental curve: solid line; numerical curve: dashed line. (c) Real part of the displacement field of the piezoelectric transducer under harmonic excitation. (transverse mode at 250 kHz)

In order to confirm this behavior, the impedance of these transducers was measured using a HP 4194A network analyzer and then computed by the finite element method (FEM). Indeed, electrical impedance as a function of frequency is a suitable indicator of resonance modes. Moreover, the computation of the displacements fields by the finite element code enables one to classify these resonance modes.

An illustration of the impedance results in the range 150–450 kHz is presented in Figure 3(b). A very good agreement between the experimental testing and the numerical analysis is observed in this graph. From these curves, one natural vibration mode is clearly visible. Figure 3.c shows the real displacement field of the piezoelectric under excitation at 250 kHz frequency, and it allows one to identify it as a transverse mode. The coupling coefficient of this mode could be determined using the following equation [15]:

$$k_e = \sqrt{1 - \left(f_r \middle/ f_a \right)} \tag{1}$$

where f_r and f_a are the resonance and anti-resonance frequencies respectively, of the vibrational mode. This preliminary study, therefore, confirmed the ability to use the transverse resonance of the transducer to excite ultrasonic waves with a satisfactory electromechanical coupling since k_e was equal to 43%.

Once the transducers selected, the second task dealt with the choice of their location which is extremely important for successful damage detection. Attenuation measurements allowed knowing how far stimulations could be transmitted with a sufficient signal to noise ratio, i.e. more than 30dB. The results showed that for wave propagation parallel to the stringer and a frequency of 250 kHz, which was the working frequency chosen in active mode, stimulations could be propagated for a distance of 80 cm.

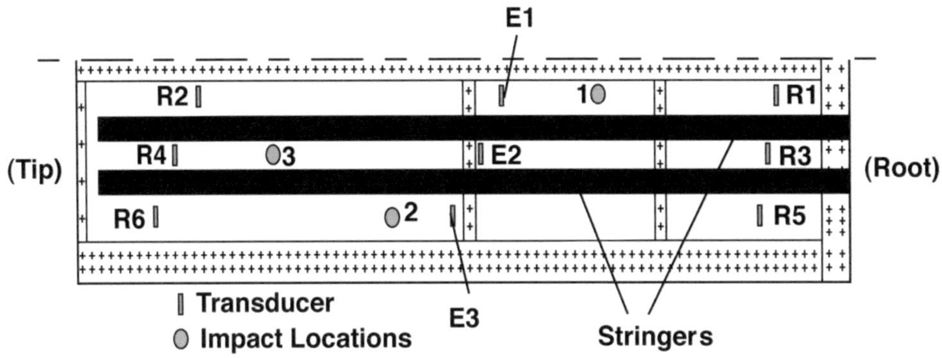

Figure 4. Schematic diagram of the health monitoring set-up –view from underside of the skin

According to these results, nine rectangular piezoelectric elements E_i *(i=1 to 3)* and R_j *(j=1 to 6)* were bonded on the inner surface of the skin as shown in Figure 4. This figure indicates also which of the transducers were used in the actuation *(E)* and which in the sensing *(R)* mode respectively. Moreover, the structure monitoring was divided in six zones using the active and passive system. Generation of a pure single ultrasonic wave between emitters and receivers in active mode was not always possible due to the complexity of the monitored structure as explained in section 2.4, but the integrity of each zone could be monitored. In passive mode, the AE events during each impact were readily detected by the transducers R_j *(j=1 to 6)*.

The equipment used for the instrumentation consisted in two digital oscilloscopes (Lecroy Type LT344) of four channels and three arbitrary calibrated generator functions (HP 33120A). In active mode, the three arbitrary generator functions delivered at discrete time intervals a 250 kHz, 5 –cycle tone burst modified by a Hanning window envelope to the emitters. All the signals from the sensors, i.e. stimulations or AE events were recorded from the digital oscilloscopes and transferred via a GPIB bus to a computer for signal processing.

2.4. Lamb waves system calibration

The laminate nature of fibre reinforced composite materials means that structures can be readily approximated to plate-like structures. As such it can be assumed that AE signals measured in passive mode and stimulations produced in active mode by the piezoelectric transducers will be propagating as Lamb waves adding a further level of complexity. Lamb waves are elastic perturbations propagating in a solid plate with free boundaries, for which the displacements correspond to various basic propagation modes, with symmetrical and antisymmetrical vibrations. For a given plate thickness *d* and acoustic frequency *f*, there exists a finite number of such propagation modes specified by their phase velocities. A complete description of such propagation characteristics for plates is normally given in the form of a set of dispersion curves, illustrating the plate-mode phase velocity as a function of the frequency–thickness product [16]. Each curve represents a specific normal mode, which is conventionally called A_0, S_0, A_1, S_1, A_2, S_2, etc. A_n denotes antisymmetrical modes and S_n denotes symmetrical ones.

For an optimal use of the active and passive health monitoring system, it was of primary importance to know the characteristics of the Lamb waves that can be propagated in the composite evaluator. In this way, preliminary tests were carried out in order to measure experimentally the velocities of Lamb wave signals as function of the location in the structure and of the frequency. The technique used to perform the analysis of propagating multimode signal was based on a two-dimensional Fourier transform described in [16-17]. Hence, for each thickness variation of the structure, a series of 64 waveforms was recorded along the longitudinal direction with an increment in the position of 1mm. Each Lamb wave response consisted of 1000 samples and the transducers chosen for these tests were the conventional surface mounted transducers (Panametrics A143-SB). The sampling serial of the experiments was 500ns. Before applying the two-dimensional Fourier Transform to the data matrix, 64 zeros and 24 zeros were padded to the end of the signal in both spatial and temporal domains respectively, in order to smooth the results. This method enables to describe the amplitude of Lamb wave signal as function of the frequency and of the wave number.

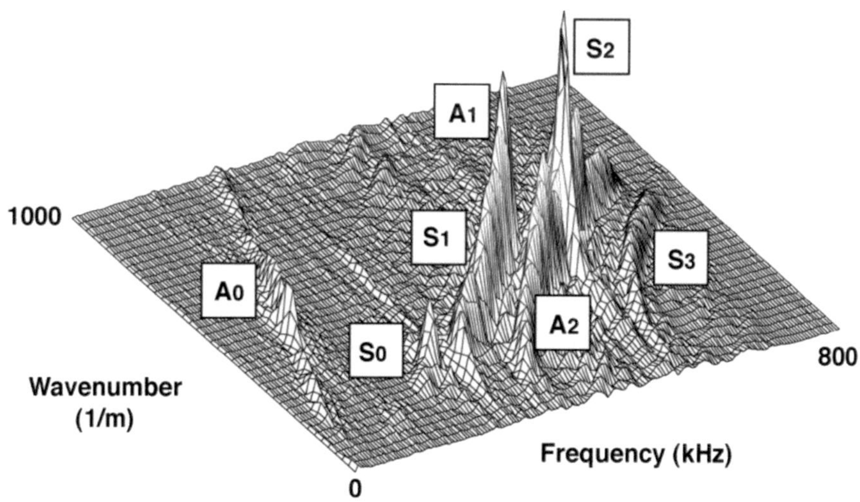

Figure 5. Measurements of the experimental Lamb wave numbers

Figure 5 illustrates an example of the measurements performed from position 800 mm to the position 1400 mm as indicated on Figure 14 in Appendix A. It can be noticed that seven different Lamb modes are propagated between 0 and 800 kHz. In addition, the theoretical phase velocities for three different positions were then computed using the formalism in [18] and compared to the phase velocity experimental measurements deduced from Figure 5. It is shown that the experimental measurements and theoretical results are in good agreement for the first modes (see Figure 6), which is sufficient since the chosen excitation in active mode was 250 kHz and the frequency range of AE events in passive mode did not exceed this value (see section 3.2). This comparative study was performed for each composite plate lay-up leading to satisfactory results.

The computing of the dispersion curves allowed estimating the Lamb modes that could appear in each region for the active mode. Hence, they were only two modes the S_0 and A_0 modes that could exist between the tip and the middle of the panel, while four modes, the S_0, A_0, S_1 and A_1, were expected between the middle and the root of the panel at 250 kHz. Moreover, the amplitudes of the A_0 and A_1 modes, which have wave structures where out-of-plane displacements are dominant at the surface, were relatively small at 250 kHz, because the transducers used in this study were more sensitive to in-plane displacements rather than out-of-plane displacements.

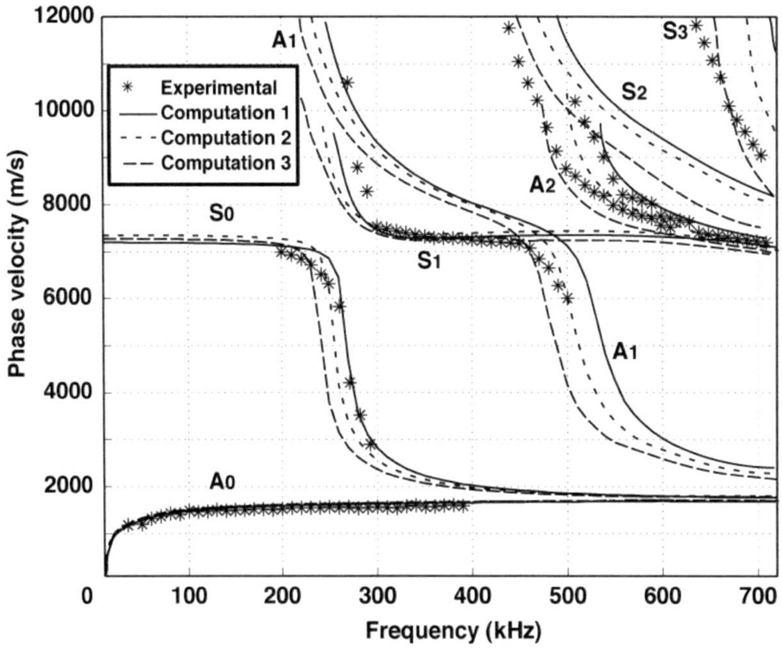

Figure 6. Experimental and theoretical phase velocity determination; Computations 1, 2, 3: thicknesses of 4.5 mm, 4.75 mm and 5 mm respectively

2.5. Damage detection and location procedure

The protocol adopted for the damage detection and location was the following: the proposed health monitoring technique started with the collection of stimulations between each pair of emitter E_i *(i=1 to 3)* and receiver R_j *(j=1 to 6)* for the healthy structure. The series of impacts were then performed at three different locations as indicated on Figure 4 by circles and the AE events were readily detected by the transducers elements R_j *(j=1 to 6)*. After each impact, new collections of stimulations signals were recorded for comparison to the initial ones. As explained in section 2.2, the skin was also examined using the manually C-scan ultrasonic system to provide a rapid assessment of damage state and to allow its severity analysis.

Interpreting the location and severity of damage by using directly time domain signals is often difficult. Consequently, it was decided to implement wavelet transforms for analysing the signals obtained in both passive and active modes and extracting the useful information. In the wavelet analysis, basis functions used were small waves of different scales located at different times of sensor signals that transform the signal to time-frequency scales. Concentration of the signal energy on the time-frequency plane was therefore obtained in terms of amplitude of wavelet coefficient at individual frequency scales.

There are many types of basis wavelet functions such as the Shannon wavelet, Morlet Wavelet, Meyer wavelet, Mexican hat wavelet, Gabor wavelet, etc. Out of these basis wavelet functions, the Gabor wavelet function was adopted in this study since it is known to provide the best time-frequency resolution [19-20]. This wavelet $\psi_g(t)$ is expressed by the following equation :

$$\psi_g(t) = \frac{1}{\sqrt[4]{\pi}}\sqrt{\frac{\omega_0}{\gamma}}\, e^{-\frac{(\omega_0/\gamma)^2}{2}t^2} e^{i\omega_0 t} \tag{2}$$

and its Fourier transform is :

$$\hat{\psi}_g(\omega) = \frac{\sqrt{2\pi}}{\sqrt[4]{\pi}}\sqrt{\frac{\gamma}{\omega_0}}\, e^{-\frac{(\gamma/\omega_0)^2}{2}(\omega-\omega_0)^2} \tag{3}$$

where $f_0 = \omega_0/2\pi$ is the central frequency of the Gabor wavelet and $\gamma = \pi\sqrt{2/\ln 2} \approx 5.336$ a shape control parameter. The Gabor function (see eq.2) may be considered as a Gaussian Function centred at $t = 0$ and its Fourier transform (see eq.3) centred at $\omega = \omega_0$. Using the Gabor function as mother wavelet, the continuous wavelet transform (CWT) of an harmonic waveform $u(x,t)$ is defined as [19-20] :

$$WT(a,b) = \frac{1}{\sqrt{a}}\int_{-\infty}^{+\infty} u(x,t)\,\psi_g^*\left(\frac{t-b}{a}\right)dt \tag{4}$$

where the continuous variables a and b are the scale and translation parameters of the Gabor function respectively, its bandwidth being proportional to $1/a$ (see Fig. 7). The function $WT(a,b)$ using the Gabor wavelet thus represents the time-frequency component of $u(x,t)$ around $t = b$ and $\omega = \omega_0/a$. By setting $\omega_0=2\pi$, we get $1/a$ equal to the frequency f. The square modulus of the wavelet transform $WT(a,b)$ is associated to the energy distribution of the signal and is also referred to as a scalogram. See [21] for the detailed analysis of the Lamb wave signal using wavelet analysis.

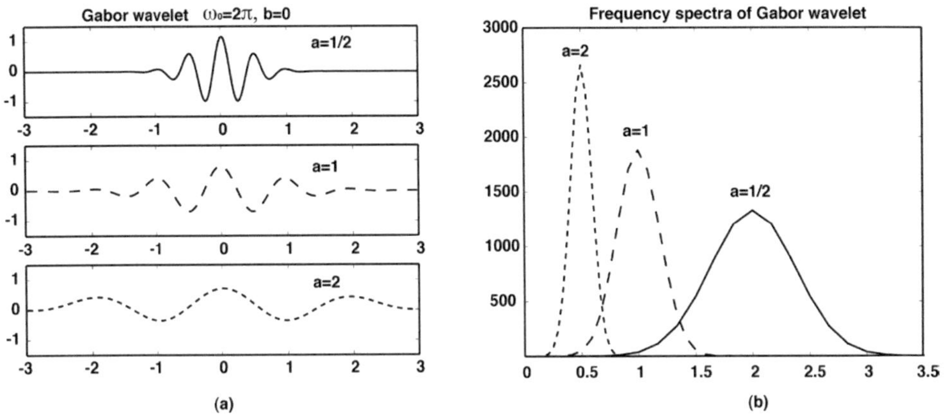

Figure 7. Gabor function at different scales (a) Real part; (b) Modulus Fourier Transform

In order to detect the presence of damage in structure, two kinds of damage sensitive features were deduced from the signals wavelet analysis: a) feature I: maximum value of the received energy in active mode (see Section 3.1) and b) feature II: maximum value of energy of high frequency AE due to damage (see section 3.2) in passive mode. From these features, appropriate damage indexes DI_A^n and DI_p^n which should reflect changes in the received data due to the damage were proposed as follows:

$$DI_A^n = \left| \frac{\max_{a,b\in R^2}\left(\left|WT_A^0(a,b)\right|^2\right) - \max_{a,b\in R^2}\left(\left|WT_A^n(a,b)\right|^2\right)}{\max_{a,b\in R^2}\left(\left|WT_A^0(a,b)\right|^2\right)} \right| \tag{5}$$

$$DI^n{}_P = \frac{\max_{a\in R,b\in B_2}\left(\left|WT_P^n(a,b)\right|^2\right)}{\max_{a\in R,b\in B_1}\left(\left|WT_P^n(a,b)\right|^2\right)} * \max_{k\in[1\,\mathrm{to}\,n]}\left[\frac{\max_{a\in R,b\in B_1}\left(\left|WT_P^k(a,b)\right|^2\right)}{\max_{a\in R,b\in B_2}\left(\left|WT_P^k(a,b)\right|^2\right)}\right] \tag{6}$$

where the superscript n represents the nth impact, N the total number of impacts and the subscripts A and P correspond to the active and passive modes, respectively. The domains B_1 and B_2 refer to the selected time ranges in the time-frequency plane, B_1 corresponds to the duration of the impact whereas B_2 is associated to the duration of the damage AE events (see sections 2.2 and 3.2). $WT_P{}^n$ are the wavelets coefficients computed from AE signals obtained for each successive n^{th} impact, whereas $WT_A{}^0$ and $WT_A{}^n$ correspond to wavelets coefficients computed from active signals obtained for the healthy structure and after each nth impact, respectively.

Notice that damage indexes $DI_A{}^n$ and $DI_p{}^n$ are theoretically equal to zero when no damage is observed. Moreover, threshold values TR_A and TR_P can be obtained from these indexes (see section 3.3). If the damage indexes are larger than the threshold values, damage is detected and its location can be performed. Then, the numbering of damage indexes is reset to zero to allow the detection of damage at other locations.

Locating an AE source or an impact is an inverse problem. If we assume in a first approximation that the group velocity Vg of the AE waves measured from the sensors R_j *(j=1 to 6)* is constant in the structure, the coordinates (x_n, y_n) of the nth impact source can be determined by solving the following set of non-linear equations:

$$(x_{Rj}-x_n)^2+(y_{Rj}-y_n)^2-\left[\left(t_m \pm \Delta t_{mj}\right)V_g\right]^2=0 \tag{7}$$

Where (x_{Rj}, y_{Rj}) correspond to the coordinates of the jth sensor, t_m is the travel time required to reach the sensor R_m *(m=1 to 6)* and Δt_{mj} are the time differences between sensors R_m and R_j.

For solving this set of non-linear equations with three unknown variables $[x_n, y_n, t_m]$, the method adopted was to combine a Newton's method with an unconstrained optimization in order to ensure the algorithm convergence [22]. Moreover, differences Δt_{mj} in time-frequency wavelet scalograms were used for accurately measuring the arrival time differences.

3. Experimental results and discussion

3.1. Baseline signals in active mode

In active mode, baseline signals were measured for the healthy structure between each pair of emitter E_i *(i=1 to 3)* and receiver R_j *(j=1 to 6)*. Each Lamb wave response consisted of 2000 samples and the sampling frequency was equal to 2MHz. Figure 8 illustrates Lamb wave responses received on the root and on the tip of the panel, respectively. Figure 8(a) shows the Lamb wave response on the transducer R_1 when transducer element E_1 was excited, whereas Figure 8(b) represents the Lamb wave response on the transducer R_6 when transducer element E_3 was excited. In both cases, the energy spectral density of the signal was concentrated around 250 kHz which was the frequency chosen to excite the emitters at their transverse resonance.

From measurements of the time of flights (or arrival times) tS_1 and tS_0 of each mode, the first wave packet of figure 8(a) was identified as the combination of the S_1 and S_0 modes. Since the S_0 mode was excited in a dispersive frequency region, which meant a time spread of the Lamb mode, the amplitude of the S_1 mode was dominant. Similar waveforms were obtained on the root of the panel, i.e. for transducers R_3 and R_5 after excitation of the transducers E_2 and E_3, respectively.

In contrary, while analysing Figure 8(b), it was concluded that the propagated signal corresponded mainly to the propagation of the S_0 mode. Similar waveforms were obtained on the tip of the panel, i.e. for transducers R_2 and R_4, after excitation of the transducers E_1 and E_2, respectively.

In both cases, wavelet analysis was applied in order to estimate the maximum of transmitted energy. The study of the arrival time of the maximum spectral energy density confirmed the identification of the Lamb modes made previously. Measurements performed for the healthy structure and after each impact in active mode will be used in the following to compute the damage index DI_A^n proposed eq.5 and therefore evaluate the sensitivity of these Lamb modes to the damage.

3.2. AE signals preliminary study and processing

Received AE signals for a 6J impact at the location 1 are shown in Figure 9. Each AE signal consisted of 20000 samples and the sampling frequency was equal to 1MHz. Among all received signals, the largest signal arriving earliest in time was that from the sensor R_1 [see figure 9(a)] that was nearest the impact location. As the propagation distances from the impact location to different sensors R_j *(j=1 to 6)* varied significantly, the shapes of the signals recorded by different sensors looked significantly different due to dispersion and attenuation on the stringer.

Since the family of nonlinear equations contained only three unknowns, only three sensors were required to obtain these arrival time differences. Sensor selection was then carried out taking the three first sensors for which the value of amplitude of the time domain signal exceeded a given threshold as shown in Figure 9. It allowed choosing the three sensors with the greatest signal to noise ratio. For that case, sensors R_j *(j=1 to 3)* were selected. Since recording the arrival time differences by the threshold technique directly on the time domain signals did not give accurate results, it was also necessary to use the wavelet scalogram technique for accurately extracting these arrival time differences.

Figure 10 illustrates the procedure for extracting the arrival time differences at the frequency of interest. Firstly, a high pass filter with a cut-off frequency of 15 kHz was applied to all the AE signals in order to privilege the analysis of high frequencies. Indeed, wavelet transform results in better time resolution at higher frequencies. Secondly, the scalogram for each AE signal was performed and represented in contour plot. Thirdly the maximum of the energy spectral density at the frequency of interest was used to estimate the differences of arrival time between the sensors [23]. A preliminary study was therefore done for this first impact of 6J at

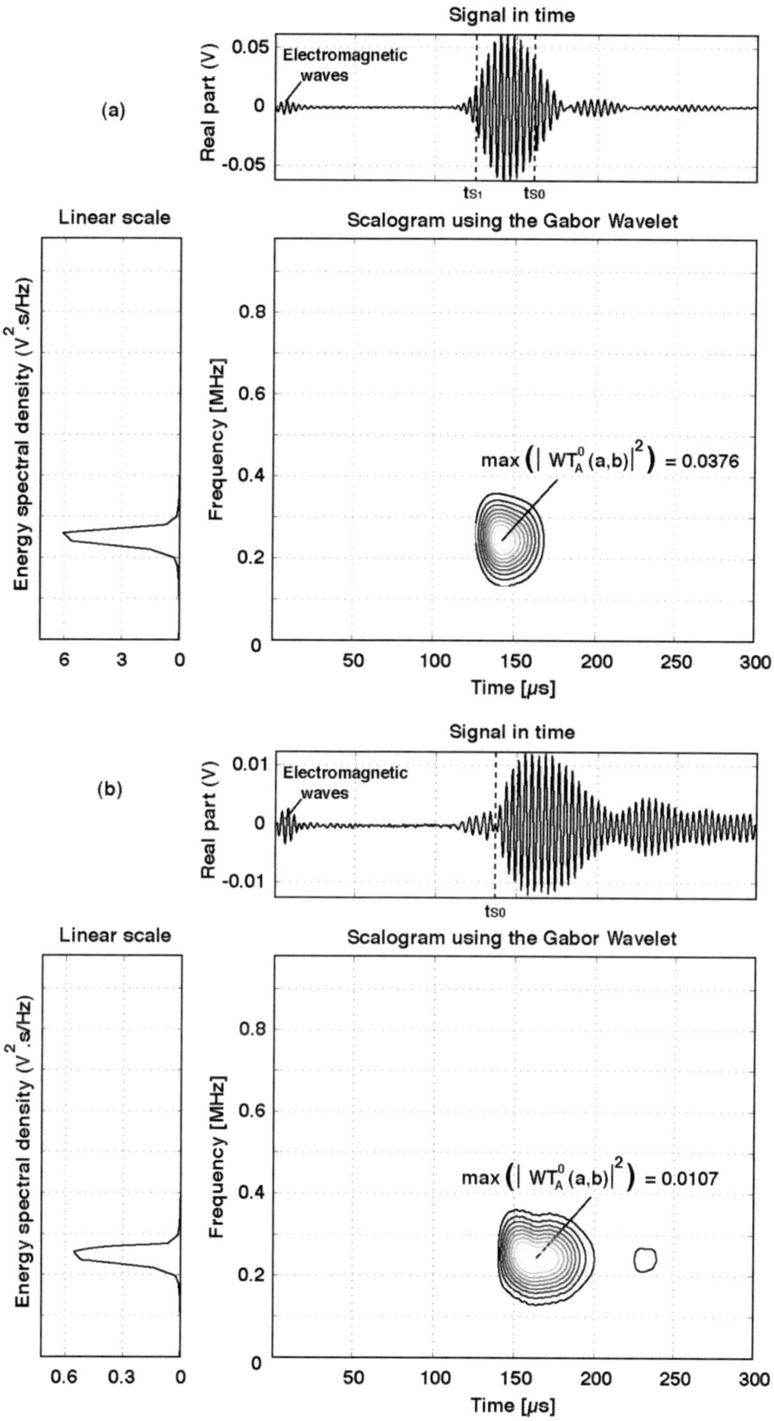

Figure 8. Lamb wave Response and its scalogram (a) at receiver R_1 for the undamaged condition, (b) at receiver R_6 for the undamaged condition

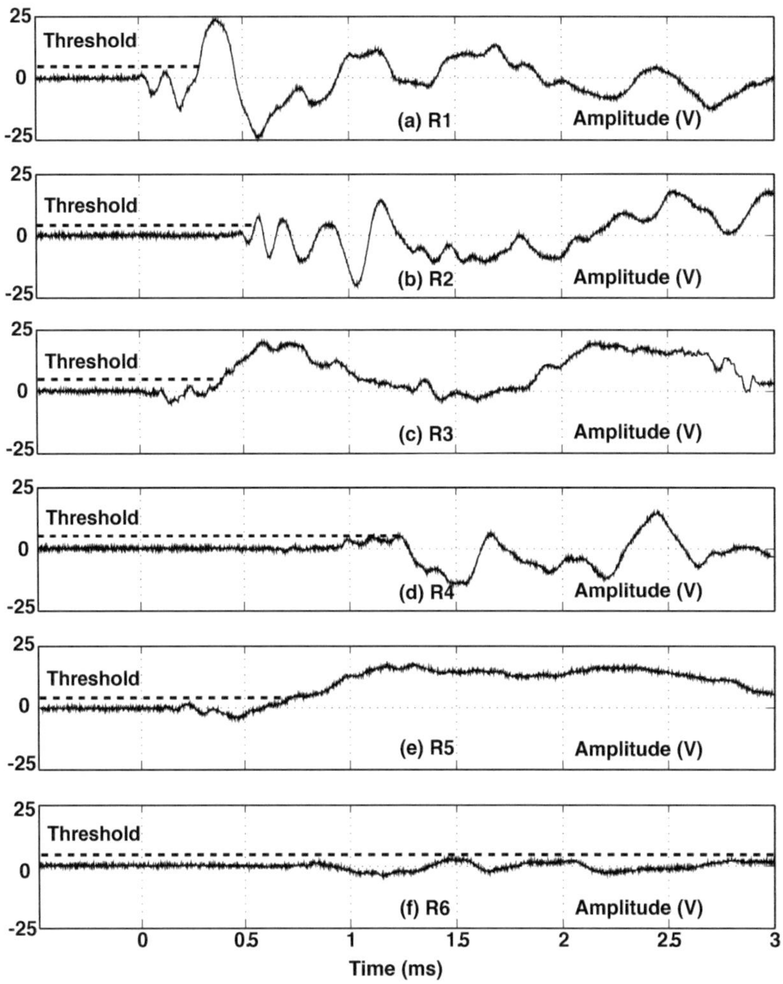

Figure 9. AE data for a 6J-energy impact at location 1 recorded from the sensors :(a) R_1; (b) R_2; (c) R_3; (d) R_4, (e) R_5 and (f) R_6

location 1 to choose the frequency of interest for all the successive impacts and also to estimate the group velocity of the Lamb mode with the dominant energy in the AE signal.

From the plots of Figure 9, it was clear that the exact arrival time of the weak S_0 mode could not be determined since this mode was hidden in the low level noise present in the time history plot. In contrary, the projection (see Figure 10) on the time domain of the ridge around the instantaneous frequency 30 kHz corresponded to the time of arrival of the A_0 mode.

This frequency was chosen since the wavelet analysis showed that the A_0 mode was relatively few dispersive around 30 kHz and was therefore independent from the thickness variation of the structure. Moreover, although signal to noise ratio was higher at lower frequencies, single-velocity arrival times was found to be robust in the presence of electronic noise even for low signal-to-noise ratios. This confirmed the results given in [24].

http://dx.doi.org/10.5772/54581

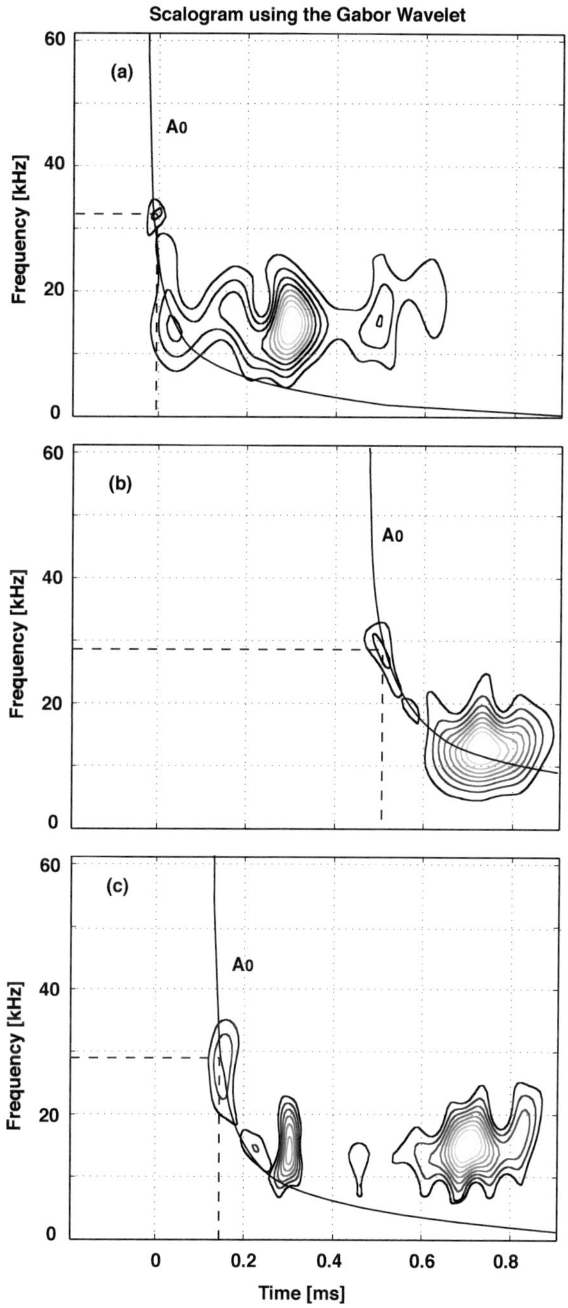

Figure 10. Scalogram of filtered AE data for a 6J-energy impact at location 1 recorded from the sensors :(a) R_1; (b) R_2; (c) R_3

Knowing the location of this first impact a priori, the group velocity was deduced from the difference in arrival time between sensors using eq. 7. The group velocity was equal approximately to 1800m/s, which was in good agreement with the group velocity calculated from the phase velocity measurements of section 2.4. According to the quasi-isotropic nature of the composite plate, the group velocity was considered as constant in all the directions of the plate. Another advantage of using the A_0 mode for this range of frequencies was that the responses of the finite-size sensors converged to that of the point sensors [25] and therefore, the size and shape of the sensors did not influence the measurements.

It was observed during the test that due to severe attenuation of the stringer, the scalogram maxima coefficients in such sensors resulted slightly different. However, the associated frequencies were approximately the same (within a band of 3 kHz) with respect to the nominal value of 30 kHz. This means that the arrival time evaluation error due to frequency shift was negligible.

Figure 11(a) andFigure 11(b) represent the AE signals recorded by the sensor R_1 during impact at location 1 with energy levels equal to 6J and 40J, respectively. After comparison, it could be noticed from the time domain signals that the AE signal recorded for a 40J impact showed the emergence of a second wave packet composed of high frequencies. This result was confirmed using the wavelet analysis, since frequencies until 80 kHz were observed. This meant that the first wave packet corresponded to the impact duration whereas the second wave packet could be related to the damage emergence. This phenomenon already observed on the force sensor (see section 2.2) will be used in the following for the damage index evaluation DI_p^n proposed eq.6. Moreover, this figure allowed estimating the values of the domains *B1* and *B2* denoted in this equation.

3.3. Damage index

In order to facilitate the damage detection and to avoid false alarms, the damage indices defined in eq.5 and eq.6 were calculated and plotted on a two-dimensional damage feature (2-D DF) space. Results for the series of impacts at the three locations are presented in Figures 12(a), 12(b) and 12(c) respectively. For each case, solely the results of the three most sensitive sensors were plotted. A C-scan analysis was therewith proposed for each figure when damage occurred. The amplitude values for these C-scans were represented with an arbitrarily determined colour assignment. According to the "traffic light effect", small indications below the tolerance limit were displayed blue, critical indications were displayed pink, and indications exceeding the tolerance limit were displayed yellow and white.

Considering the results of Figure 12(a), it could be observed first that the damage indexes for non-damaging impact were always lower than the threshold values TR_A =0.2 and TR_P =0.2. On other hand, it could be also noticed that the three sensors R_1, R_2 and R_3 were mostly affected by the impact with energy level of 40J. Indeed, measured values for the passive damage index DI_p were greater than 0.4. As shown in Figure 11(b), these variations corresponded to the emergence in the AE signal of a second wave packet composed of high frequencies when the damage occurred. For this case, the C-scan showed a damage size of approximately 2500 mm^2.

After analyzing the values of active damage index DI_A from Figure 12(a), only the sensor R_1 seemed perturbed since a value greater than 0.5 was measured. From this information, it was possible to delimit the zone where the damaging impact occurred, i.e. the zone monitored by the emitter E_1 and the receiver R_1 in active mode. Another essential outcome was that this variation of the active damage index revealed a large sensitivity of the Lamb mode S_1 to the damage.

Besides, similar conclusions could be made for the impacts at location 2 (see Figure 12(b)), although only two impacts were performed. In this case, the results demonstrated again a strong sensitivity of the three sensors R_6, R_5 and R_4 to the damaging impact at 40J, as their passive damage indexes DI_p were greater than 0.4. Moreover, a smaller variation of damage index DI_A than in the preceding case was observed for the sensor R_6. This revealed a lower sensitivity of the Lamb mode S_0 than the Lamb mode S_1 to the damage. Finally, results allowed estimating the zone of the damaging impact, i.e. between the emitter E_3 and the receiver R_6.

As expected, similar results were also obtained for the series of impacts at location 3. In fact, the three sensors R_6, R_4 and R_2 demonstrated a great sensitivity to the damaging impact with energy level equal to 40J. Nevertheless, Figure 12(c) enabled also to determine the sensitivity limit of the used health monitoring since a low damaging impact (energy with a level equal to 35J) was tested. For this case, a damage size of 150 mm^2 was measured, but the health monitoring system seemed to be little perturbed, particularly in active mode where the damage index measured DI_A was lower than 0.1. A solution to improve the active system sensitivity would be to work at higher frequencies, but this would require an increase in the number of transducers due to a larger attenuation in the composite at these frequencies.

Taken together, these results demonstrate the feasibility of the proposed health monitoring system to detect damaging impact despite the complexity of the structure. Among the possible ways for improvement of this monitoring and especially to have information also on the severity and characteristic of the damage, one solution would be to develop models in order to better understand the effects of interaction of Lamb modes with damage [26], but also to be able to characterize the sources of damage from acoustic signals [27]. Modeling studies are already proposed for composite structures but very few works are concerned with the case of structures composed of stiffeners [28].

3.4. Damage location

Fig. 13 shows the source location results for all damaging impacts. The real impact positions are plotted with a circle along with the calculated location using the wavelet transform method. Damaging impact locations were estimated each time using the three most sensitive sensors in passive mode (see section 3.2). The data comparisons show good agreement. Table 1 summary these results together with errors. The error was expressed by the following formula :

$$\psi = \sqrt{\left(x_n^r - x_n^c\right)^2 + \left(y_n^r - y_n^c\right)^2} \tag{8}$$

Where (x^r_n, y^r_n) are the coordinates of the real impact position and (x^c_n, y^c_n) the coordinates of the impact location using the algorithm reported in section 2.5. The errors here could be attributed

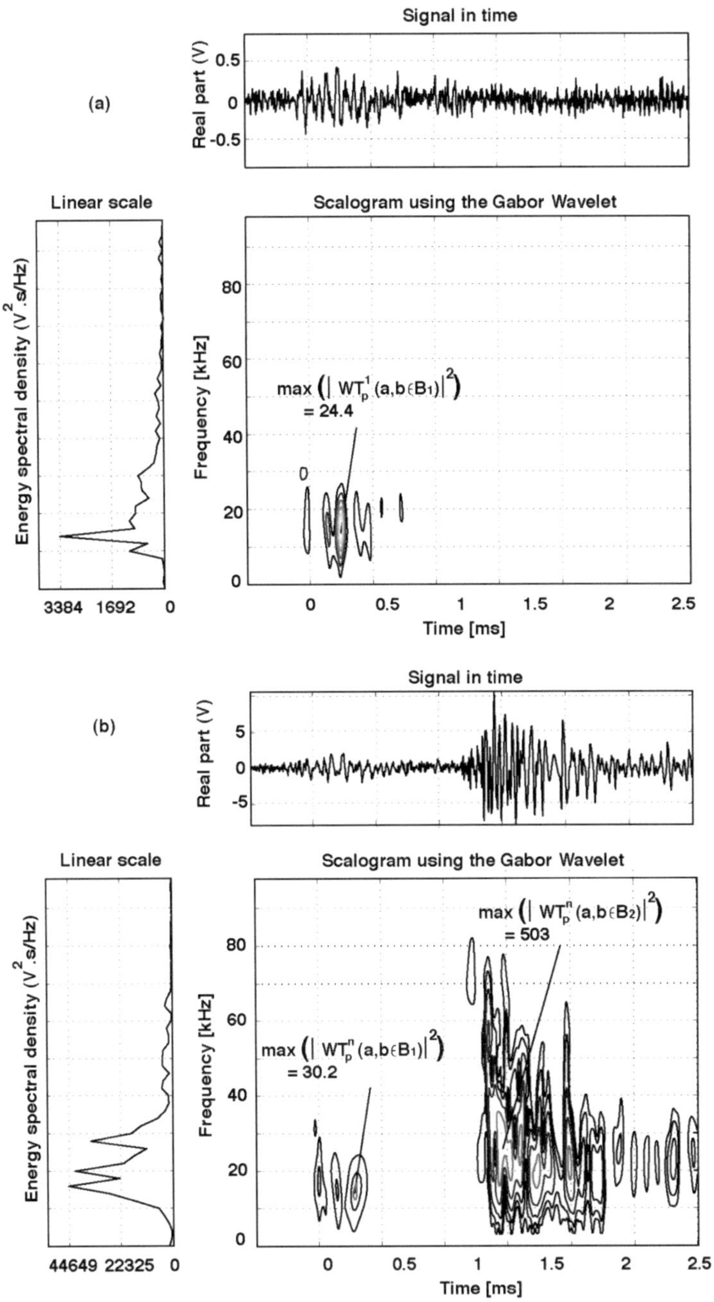

Figure 11. Scalogram of filtered AE data (a) for a 6J and (b) 40J energy impact at location 1 recorded from the sensor R1

to material constants used to calculate the theoretical curves, measurement errors in the placement of sensors and location of the impinger. More likely, this can be due to the presence of the stringers which were not taken into account in the estimation of the group velocity.

http://dx.doi.org/10.5772/54581

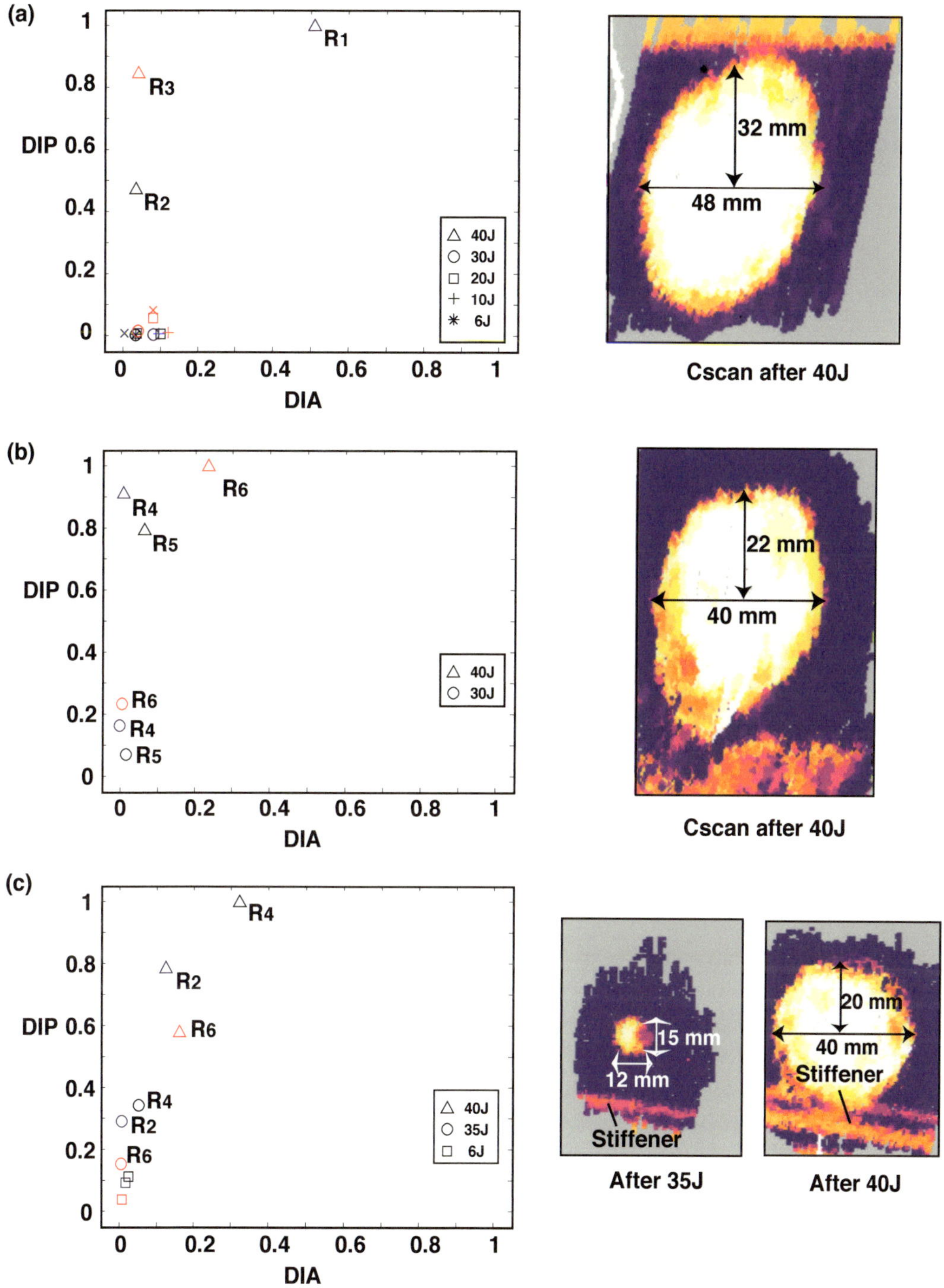

Figure 12. Damage feature space for the series of impacts and the C-scan analysis (a) at location 1, (b) at location 2, (c) at location 3

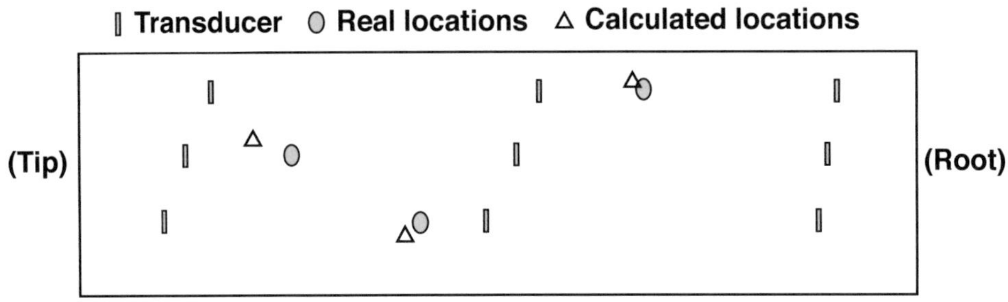

Figure 13. Source location results.

	Impact location 1	Impact location 2	Impact location 3
x-coordinate source location (from algorithm) [mm]	1187	700	375
x-coordinate source location (real value) [mm]	1150	737	462
y-coordinate source location (from algorithm) [mm]	332	95	237
y-coordinate source location (from algorithm) [mm]	314	114	218
Location error [mm]	41	42	89

Table 1. Impacts positions and errors

4. Conclusions

In this paper, experimental investigations were developed to demonstrate the feasibility of using a same piezoelectric Health monitoring system based on both passive and active mode for the inspection of a large and complex structure submitted to impacts. On one hand, the consideration of the AE signatures during impact loadings was of interest as the information retrieved can allowed one to estimate impact location and second to detect a damage occurrence. On the other hand, the experimental results presented here for the active monitoring system revealed also a large sensitivity of the generated Lamb modes to the damage and allowed to confirm the passive diagnostic. Damages indexes were also developed using wavelet analysis and allowed to avoid false alarm, despite the false detection probability could not be measured. Finally, the goal of detecting damaging due to impacts in a complex structure with a minimum of transducers was achieved since only 9 transducers were used for a surface of 1800*380mm^2. More work on the active monitoring and more precisely on the Lamb wave interaction with damage is now required in order to extract the information related to the severity of damage.

http://dx.doi.org/10.5772/54581

Appendix A: Layup and geometry of the structure

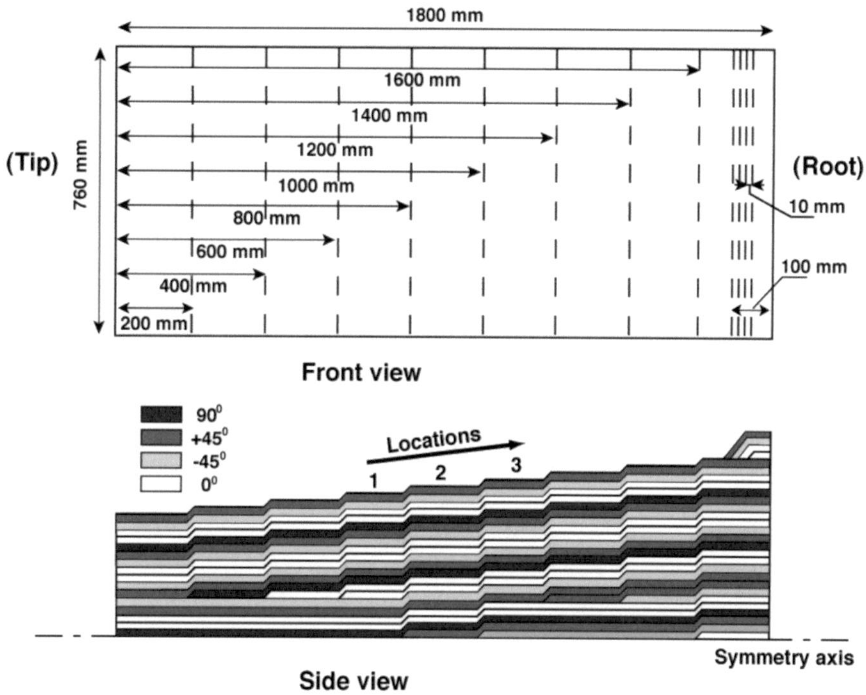

Figure 14. Front and side view of the wing-box structure

Appendix B: Material Data for the 913C-HTA

The skins were made from 913C-HTA material by DSTL (UK). The x_3-axis was defined as the perpendicular axis to the ply-plane.

Nominal ply thickness: 0.125mm, Mass density: 1630kg/m^3, Elastic tensor: E_{11} =158 GPa, E_{22} =E_{33}=9.7 GPa, ν_{12} =ν_{13} =0.3, ν_{23} =0.6, G_{12}=G_{13}=7.4 GPa, G_{23} =3.6 GPa.

Acknowledgements

This work has been performed under the Brite/Euram Monitor (Monitoring On-line Integrated Technologies for Operational Reliability) project no. BE95/1524. The authors acknowledge the support of the Monitor team whose efforts have till date produced successful research. We would also like to thank the C.E.C. who, through their Brite Euram program, has made this research possible.

Author details

Sébastien Grondel* and Christophe Delebarre

*Address all correspondence to: sebastien.grondel@univ-valenciennes.fr

IEMN, Department OAE, IEMN, UMR CNRS 8520, Université de Valenciennes et du Hainaut Cambrésis, Le Mont Houy, Valenciennes, France

References

[1] Staszewsky W. J., Boller C., Tomlinson G., editors. Health monitoring of aerospace Structures – Smart Sensor technologies and Signal processing. Editions John Wiley & Sons; 2004.

[2] Grondel S., Delebarre C., Assaad J., Dupuis J.-P., Reithler L. Fatigue crack monitoring of riveted aluminium strap joints by Lamb wave analysis and acoustic emission measurement techniques. NDT & E International 2002; 35 137-46.

[3] Martin T., Jones A., Read I., Murray S., Haynes D., Lloyd P., Foote P., Noble R., Tunnicliffe D. Structural health monitoring of a carbon fibre structure using low profile piezoelectric, optical and MEMS sensors. Key Engineering Materials 2001; 204-205 371-82.

[4] Eaton M. J., Pullin R., Holford K. M. Acoustic emission source location in composite materials using Delta T Mapping. Composites: Part A 2012; 43 856-863.

[5] Tsutsui H., Kawamata A., Sanda T., Takeda N., Detection of impact damage of stiffened composite panels using embedded small-diameter optical fibers. Smart Mater. Struct. 2004; 13 1284.

[6] Bocherens E., Bourasseau S., Dewynter-Marty V., Berenger H. Damage Detection in a radome sandwich material with embedded fiber optic sensors. Smart. Mater. Struct. 2000; 9 310-315.

[7] Cawley P., Adams R. D. The location of defects in structures from measurements of natural frequencies. Journal of Strain Analysis 1979; 14 49-57.

[8] Grondel S., Paget C. A., Delebarre C., Assaad J. and Levin K. Design of optimal configuration for generating A0 Lamb mode in a composite plate using piezoceramic transducers. J. Acoust. Soc. Am. 2002; 112 84-90.

[9] Paget C. A., Grondel S., Levin K., Delebarre C. Damage assessment in composites by Lamb waves and wavelets coefficients. Smart. Mater. Struct. 2003; 12 393-402.

[10] Choi K., Keilers C. H. Jr, Chang F. K. Impact damage detection in composite structures using distributed piezoceramics. AIAA/ASME/ASCE/AHS/ASC Structures, Structural Dynamics, and Materials Conf. 1994; 1 118-24.

[11] Grondel S., Assaad J., Delebarre C., Moulin E. Health monitoring of a composite wingbox structure. Ultrasonics 2004; 42 819-24.

[12] Staszewski W. J., Buderath M. Vibro-Acousto-Ultrasonics for Fatigue Crack Detection and Monitoring in Aircraft Components. In: Boller C., Staszewski W. (eds.): Proceedings of the 2nd European Workshop on Structural Health Monitoring. DEStech Publication Inc.; 2004.

[13] Lemistre M., Balageas D. A new concept for structural health monitoring applied to composite material. In: Balageas D. (ed.): Proceedings of the first European Workshop on Structural Health Monitoring. DEStech Publication Inc.; 2002.

[14] An Y. -K., Sohn H. Integrated impedance and guided wave based damage detection. Mechanical Systems and Signal Processing 2012; 28 50-62.

[15] Moulin E., Assaad J., Delebarre C. Piezoelectric transducer embedded in a composite plate: application to Lamb wave generation. J. Appl. Phys. 1997; 82 (5) 2049–2055.

[16] Grondel S., Assaad J., Delebarre C., Blanquet P., Moulin E. The propagation of Lamb waves in multi-layered plates: phase velocity measurements. Meas. Sci. Technol. 1999; 10 348.

[17] Alleyne D., Cawley P. A two –dimensional Fourier transform method for the measurement of propagating multimode signals. J. Acoust. Soc. Am. 1991; 89 1159-68.

[18] Nayfeh A. H., Chimenti D. E. The general problem of elastic waves propagation in multilayered anisotropic media. J. Acoust. Soc. Am. 1991; 89 1521-31.

[19] Chui C.K. An introduction to wavelets. Editions San Diego, CA : Academic Press; 1992.

[20] Mallat S.A. Wavelet Tour of Signal Processing. Editions San Diego, CA : Academic Press; 1998.

[21] Jeong H., Jang Y.-S., Wavelet analysis of plate wave propagation in composite laminates. Composite Structures. 2000; 49 443-450.

[22] Dennis J. E., Schnabel R. B. , editors, Numerical Methods for Unconstrained Optimization and Non Linear Equations, editions SIAM; 1996.

[23] Yamada H., Mizutani Y., Nishino H., Takemoto M., Ono K. Lamb Wave Source Location of Impact on Anisotropic Plates. Journal of Acoustic Emission 2000; 18 51-60.

[24] Hamstad H., O' Gallagher A., Gary J. A wavelet transform applied to acoustic emission signals. Journal of Acoustic Emission 2002; 20 62-82.

[25] Di Scalea F. L., Matt H., Bartoli I., Srivastava A., Park G., Farrar C. The fundamental response of piezoelectric guided-wave sensors and applications to damage and im-

pact location. In: Proceedings of the 3rd European Workshop on Structural Health Monitoring. DEStech Publication Inc.; 2006.

[26] Benmeddour F., Grondel S, Assaad J, Moulin E. Study of the fundamental Lamb modes interaction with asymmetrical discontinuities NDT & E International 2008; 41 330-340.

[27] Ben Khalifa W., Jezzine Karim, Grondel S., Lhemery A. Modeling of the Far-Field Acoustic Emission from a Crack Under Stress. In: Proceedings of the 30th European Conference on Acoustic Emission testing and 7th International Conference on Acoustic Emission Granada 2012.

[28] Reusser R.S., D.E. Chimenti, R.A. Roberts, S.D. Holland. Guided plate wave scattering at vertical stiffeners and its effect on source location. Ultrasonics 2012; 52 687–693.

Chapter 5

Generation of a Selected Lamb Mode by Piezoceramic Transducers: Application to Nondestructive Testing of Aeronautical Structures

Farouk Benmeddour, Emmanuel Moulin,
Jamal Assaad and Sonia Djili

Additional information is available at the end of the chapter

http://dx.doi.org/10.5772/54582

1. Introduction

The motivation of the this work is highlighted by the need for the Non-Destructive Testing (NDT) of aircraft, petrochemical and naval structures. Functioning conditions of these structures and the time factor can lead to serious damage. The most appropriate NDT technique to plate-like structures seems to be guided ultrasonics Lamb waves [1]. These waves can carry out energy over long distances and have the potential to be sensitive to several types of defects. Consequently, their use allows fast and efficient inspections of industrial structures. However, the multimode and dispersive nature of Lamb modes make the interpretation of the received ultrasonic signals complex and ambiguous in presence of discontinuities. Moreover, the generation and reception of a selected Lamb mode by transducers remains a difficult task due to the complexity of guided waves.

In this work, two identical thin piezoceramic transducers are designed specifically to work in the frequency band of interest. The generation of a specific Lamb mode is ensured by the placement of these transducers at the opposite sides of the plate. The selection of the A_0 or the S_0 mode is obtained by exciting the piezoceramic transducers with in-anti-phased or in-phased signal, respectively. Then, interactions of these modes with discontinuities in aluminium plates are investigated. The interaction of Lamb waves with discontinuities has been widely analysed. Among the studied defects, one can cite holes [2], delaminations [3], vertical cracks [4], inclined cracks [5], surface defects [6], joints [7], thickness variations [8] and periodic grating [9]. Moreover, the special case of rectangular notches have been carried out by Alleyne et al [10], Lowe et al [11], Jin et al [12] and Benmeddour et al [13–15].

The Two-Dimensional Fourier Transform (2D-FT) is commonly used by researchers [3, 15, 16] to identify and quantify the existent Lamb modes. However, this technique needs spatio-temporal sampling which is time consuming and not always available from experimental measurements. In this paper, the scattered waves are acquired by means of conventional piezoelectric transducers located on the plate surfaces in front and behind the damage. The fundamental Lamb modes are separated by means of the basic arithmetic operations such as addition and subtraction. This approach allows a simple separation by using only two signals acquired at the same location on the opposite sides of the structure. Hence, the reflection and the transmission of the incident and the converted fundamental Lamb modes, when they exist, can be identified and quantified. The power reflection and transmission coefficients are then obtained with the well-known average power flow equation and the power balance is verified. Measurement results confirmed that symmetric discontinuities could not induce any mode conversion in the contrary of asymmetric discontinuities. The sensitivity of the A_0 and S_0 mode is checked.

An approximation technique of the reflection and transmission of the fundamental Lamb waves from rectangular notches is used. This technique is based on the superposition of the reflection and the transmission from a step down (start of the notch) and a step up (end of the notch). To this end, an approach based on the power reflection and transmission coefficients is proposed. These power coefficients take into account only a single echo and only the A_0 and S_0 modes are studied at a given frequency taken under the cut-off frequency of the A_1 and S_1 modes, respectively. The power coefficients are determined with the help of the finite element and the modal decomposition method. The Finite Element Method (FEM) is used to compute the displacement fields, while the modal decomposition method allows to calculate the power coefficients from these displacements at a given location on the plate surfaces. Then, this aims to compute the power reflection and transmission coefficients for a notch from those obtained for a step down and a step up. The advantages of such a study are that the power coefficients of the multiple reflections can be determined even if the structure contains different thickness variations. Moreover, it allows the study of the interaction of the fundamental Lamb waves with complex discontinuities in a fast and efficient way.

Finally, experimental measurements are compared successfully with those obtained by the numerical method.

2. The piezoelectric transducers

2.1. Applications of the piezoelectric transducers

Since several decades, piezoelectric transducers are widely used in many fields of application. Some of these fields are reported hereafter. Si-Chaib et al[17] have generated shear waves by mode conversion of a longitudinal wave by using straight ultrasonic probes coupled to acoustic delay lines. The device is used in the field of mechanical behaviour of materials and the determination of acoustic properties of porous materials. Duquennoy[18] et al have used both laser line source and piezoelectric transducers to characterise residual stresses in steel rods. They measured the velocities of Rayleigh waves. Yan et al[19] have developed a self-calibrating piezoelectric transducer for the acoustic emission application. Blomme et al[20] have designed a measurement system with air-coupled piezo-based transducers. Their work deals with coating on textile, flaws in an aluminium plate, spot welds on metallic plates, tiny air inclusion in thin castings and ultrasonic reflection on an epoxy plate with a copper layer. Martínez et al[21] have designed a prototype of segmented annular arrays to produce volumetric imaging for NDT applications. Bhalla and Soh[22] have used high frequency piezoelectric transducers to monitor reinforced concrete subjected to vibrations caused by earthquakes and underground blasts. Sun et al[23] have bounded piezoceramic patches on concrete beams to investigate the structural health monitoring.

Rathod and Mahapatra[24] have chosen a circular array of piezoelectric wafer active sensors (PWASs) to localise and identify corrosion in metallic plates. The study is based on the guided Lamb waves and an algorithm based on symmetry breaking in the signal pattern.

Other types of transducers have been also developed and studied in the literature for several applications. A brief discussion about only some types of transducers is given here. Chung and Lee [25] have fabricated focusing ultrasound transducers based on spin-coated poly(vinylidene fluoride-trifluoroethylene) copolymer films. These films can be used for high frequency wave velocity measurements and for nondestructive determination of elastic constants of thin isotropic plates. Ribichini et al [26] have used electromagnetic acoustic transducers (EMATs) for their experimental measurements on different types of steel. They show the performance of bulk shear wave generated by EMATs to investigate the physical properties of materials. Lee and Lin [27] have fabricated a miniature-conical transducer for acoustic emission measurements. Bowen et al[28] have fabricated flexible ultrasonic transducers and tested a steel rod in pulse-echo mode.

2.2. Design and characterisation

Two piezo-ceramique transducers are cut in a plate of 50×50 mm^2 with the following dimensions. The length of these transducers is chosen to be $L = 50$ mm to be able to generate a unidirectional wave. The width of $w = 6$ mm corresponds to the half of the wavelength of the A_0 mode at the chosen working frequency of 200 kHz. The thickness of the transducers is equal to 1 mm. Figure 1, depicts the dimensions of a piezo-ceramic transducer used in this work. These transducers are handled by the company Ferroperm and the used piezo-ceramic type is a soft lead zirconate titanate Pz27. This material presents a high electromechanical coupling coefficients. The mechanical characteristics of the Pz27 given by the constructor are: a density of $\rho = 7700kg/m^3$, the elastic compliances $s_{11}^E = 17 \times 10^{-12}m^2/N$, $s_{11}^D = 15 \times 10^{-12}m^2/N$, the Poisson's ratio $\nu^E = 0.39$ and the mechanical quality factor $Q_m = 80$; here, the superscripts E and D designate electrical short and open circuits, respectively. Some of the piezoelectric constants are: the coupling factor $k_{31} = 0.33$, the piezoelectric charge coefficient $d_{31} = -170 \times 10^-12C/N$ and the piezoelectric voltage coefficient $g_{31} = -11 \times 10^{-3}Vm/N$.

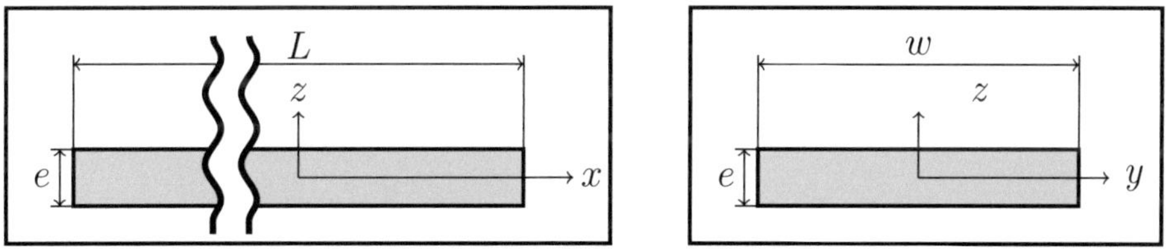

(a) (b)

Figure 1. The geometry and dimensions of the piezo-ceramic transducers (a) front view and (b) side view (arbitrary scale).

The two piezo-ceramic transducers are then characterised by a network analyser. The electrical phase and normalised module impedance are shown in figures 2a and b, respectively. It is observed that the curves of the two transducers are close each other. In addition, this can confirm the working frequency of 200 kHz and the working region near to the resonance frequency

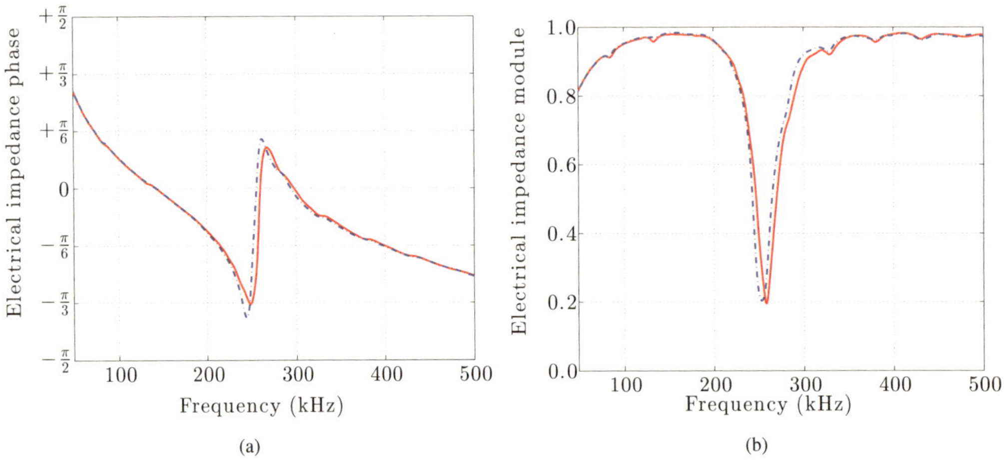

Figure 2. Electrical impedance of (a) phase and (b) module of the two used transducers (–) transducer 1 and (– ·) transducer 2.

The receiver, in this work, is an industrial transducer designed by Olympus. It is a Panametrics type with the reference A413S. Its working center frequency is 500 kHz and have a good sensitivity. The nominal element size have the dimensions of 13×25 mm^2. This transducer is also used as an emitter to characterise the aluminium plate and will be explained hereafter.

2.3. The emitter

This work aims to generate a selected guided wave in an aluminium plate. To this end, the two piezoceramic transducers are placed on the two faces of the plate in opposite position (see figure 3). It was shown [15] that the excitation of transducers with anti-phased electrical signals generates favouringly an anti-symmetrical guided wave. The excitations of these transducers with in-phased electrical signals however, produces mainly a symmetrical guided wave. The excitation of transducers is ensured by two function generators and synchronised by a pulse generator.

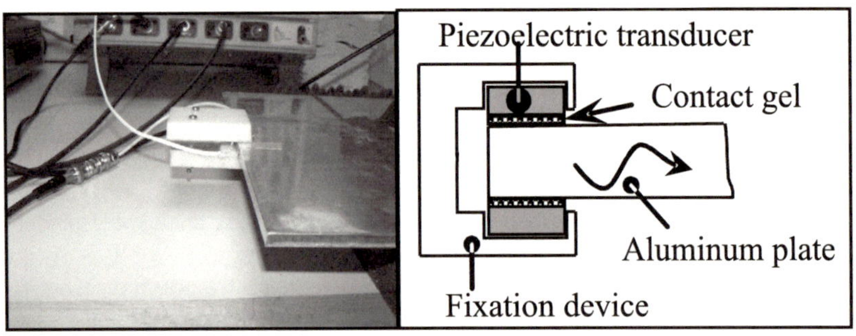

Figure 3. A photo of the piezoceramic support in-site (left) and a front view of its schematic description (right).

http://dx.doi.org/10.5772/54582

3. The experimental device

3.1. General description

As described briefly in the last section, a pulse generator (HM 8035) is used to trigger simultaneously the two arbitrary function generators (HP 33120A) and the oscilloscope (Le Croy type LT344). The function generators are able to generate a tone burst electrical signals windowed by a Hanning function to the emitter. Each of the in-phased or anti-phased signals are made of 5 sinusoidal cycles at the frequency of 200 kHz windowed by a Hanning function. The oscilloscope acquires 2000 temporal points at a sampling frequency of 10 MHz which corresponds to 0.1 μs between each point and verify the Shannon sampling theorem. Fifty acquired signals are averaged by the oscilloscope with an error less than 0.5%. Then, the result is transmitted to the computer for recording and signal processing. Furthermore, the repeatability process and the gel coupling effect are studied. To this end, the receiver is removed and replaced at the same location 20 times. Hence, an error bar can be computed for each experience.

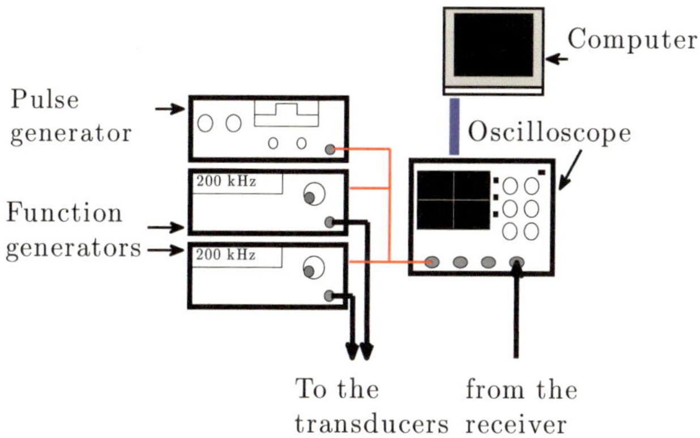

Figure 4. The scheme of the experimental device.

In what follows, aluminium plates are experimentally investigated. They have the following dimensions: 6 mm thick, 300 mm wide and 500 mm long.

3.2. Material characterisation

Before starting with experimental measurements which aim to select and generate one Lamb mode, one must characterise the plate material. For this purpose, the Panametrics transducer is used to generate a one pulse signal at a frequency of 500 kHz. Then, with a second Panametrics transducer, 256 signals are squired on a line on the surface of the plate with a 1 mm of distance between two positions. The application of a two dimensional fast Fourier transform gives rise to the dispersion curves. Figure 5, represents these dispersion curves in addition of the theoretical results computed for an aluminium plate with the following characteristics: the density $\rho = 2695$ kg·m^3, the Young's modulus $E = 72.10$ GPa and the Poisson's coefficient $\nu = 0.383$. In this figure, it is clearly shown that the chosen parameters correspond to the characteristics of the used aluminium plate. This has the advantage to identify Lamb modes by their group velocities.

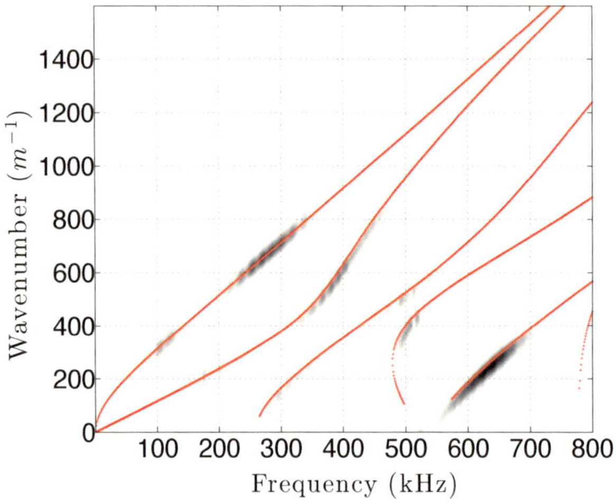

Figure 5. Experimental and analytical dispersion curves of wavenumbers vs. frequency.

3.3. Experimental measurements

By using the experimental device described and shown on figures 3 and 4, experimental measurements are carried out for in- and anti-phased electrical signals. These results are shown on figures 6a and b where the Panametrics receiver is placed at a distance of $l_1 = 185$ mm from the left edge of a healthy plate (115 mm from the right edge). The time of flight (TOF) is used to determine the group velocity of each wave packet. In fact, the used TOF is taken as the peak of the signal envelope (by using the Hilbert function) which corresponds to the center working frequency. Wave packets are clearly identified by their group velocities as the first antisymmetric (A_0) and the first symmetric (S_0) Lamb modes, respectively. In both cases, the selection of one Lamb mode is successful, which validates the experimental set-up. Since the S_0 mode has a higher group velocity than the A_0 mode, its reflection on the right edge of the plate is visible.

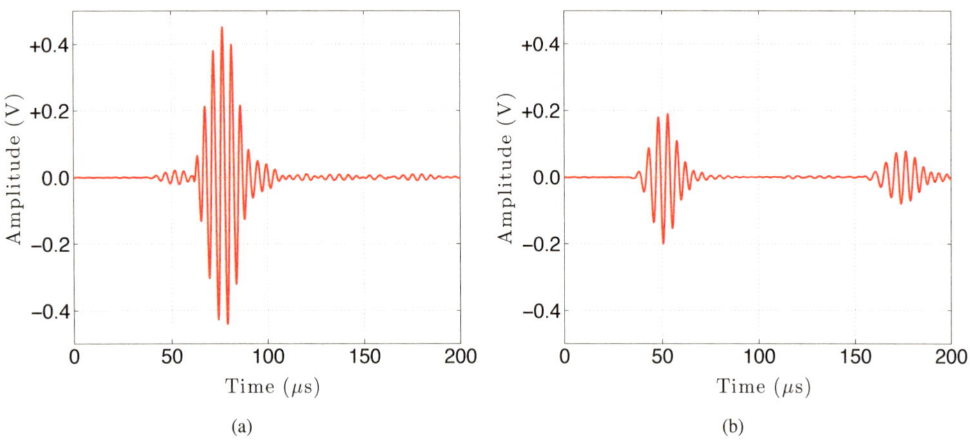

Figure 6. Measured signals when transducers are excited with (a) in-phased and (b) anti-phased electrical signals.

4. Application to the damage detection

4.1. Experimental measurements

The experimental measurements investigate the interaction of the fundamental Lamb modes A_0 and S_0 with two discontinuity kinds: symmetric and asymmetric shape-like notches. These notches are milled across the full plate width (300 mm) normal to the transmission path and have a width of 50 mm. Six notch depths are experimented and are equal to: 0.5, 1, 1.5, 2 and 2.5 mm which corresponds to the ratios p=5/6, 4/6, 3/6, 2/6 and 1/6, respectively. The desired notches width and depths are obtained with a conventional milling machine.

Owing to the existence of two modes, additional measurements and operations are conducted to separate them. To do so, signals are acquired on the two faces of the plate at exactly the same location. Then, the basic addition and subtraction operations are used. The former aims to amplify the antisymmetric mode and attenuate the symmetrical one. The latter however gives rise to enhance the symmetrical mode and to disfavour the antisymmetrical mode.

4.1.1. Calibration handling

Experimental measurements are carried out to study the receiver presence effect on the transmitted wave packets. Therefore, an identical Panametrics receiver is placed between the emitter and the conventional effective receiver. Hence, the measured signals when the A_0 and S_0 modes are launched, respectively are disturbed. In fact, the presence of the conventional transducer along the propagation path can be considered as a surface perturbation because of its dimensions ($39\times17\times15$ mm^3). Hence, the amplitude of the A_0 mode decreases to about 40%. Furthermore, its shape is modified by the multiple reflections from the edges of the receiver. Contrarily to this, the S_0 mode decreases only to about 5%.

In practice, when a fundamental Lamb mode is launched at the left edge of the plate, the incident wave packet encounter the conventional receiver. Then, the disturbed transmitted wave packet is reflected by the left edge of the discontinuity and then received. Hence, the measured reflection wave packet must be corrected by multiplying its amplitude by 40% when the A_0 mode is launched whereas the amplitude of the reflection wave packet is multiplied by only 5% when the S_0 mode is launched.

For the transmitted wave packet after the discontinuity, the transmitted amplitude does not need any correction. However, in order to take into account implicitly the dispersion, the diffraction and the attenuation effects, the incident wave packet is acquired on a healthy plate at the same location as the transmitted wave packet in a damaged plate.

4.1.2. Symmetrical notches

Figures 7a and 7b display the addition and the subtraction mean results, respectively of the electrical signals measured before a symmetrical notch when the A_0 mode is launched for a plate containing a 2.5 mm deep notch (p=1/6 of normalised thickness of the plate). By means of the flight time and the shape observation of each wave packet, no mode conversion from the incident A_0 to the S_0 mode is observed in the figure 7a. On this figure, the first and the second wave packets correspond to the incident (in) and the reflection (re_1) from the first edge of the notch. Furthermore, multiple reflections from the edges of the notch and the plate are observed. Figure 7b proves one more again that no significant mode conversion is noticed and the observed small wave packets rises mainly from the emitter.

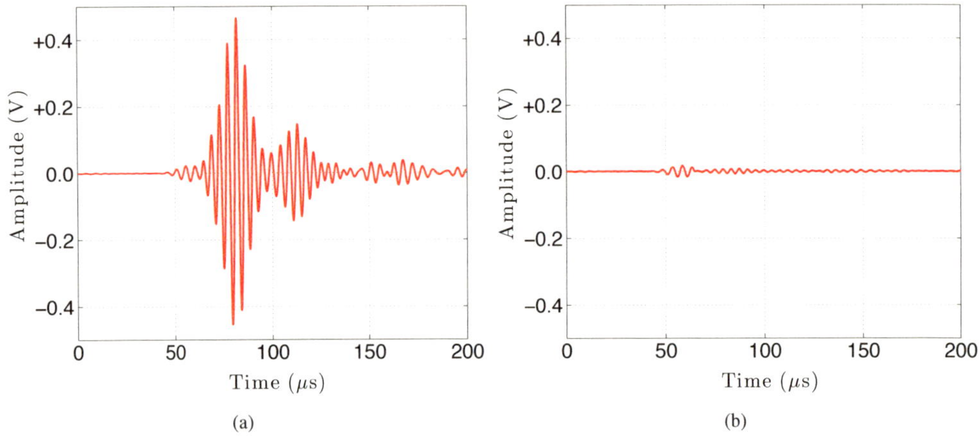

Figure 7. (a) addition and (b) subtraction mean results of the electrical signals measured before a symmetrical notch of 2.5 mm of deep (p=1/6) when the A_0 mode is launched.

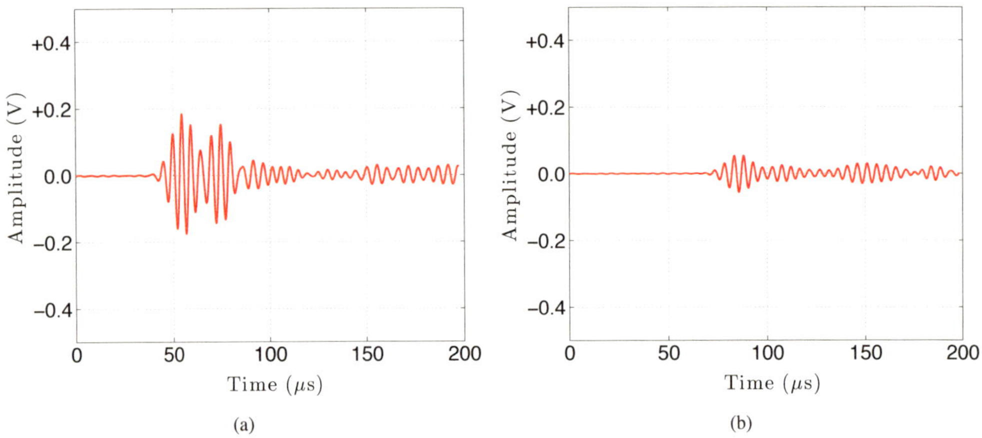

Figure 8. Subtraction mean results (a) before the notch at 185 mm and (b) after the notch at 315 mm from the left plate edge.

Figures 8a and b depict only subtraction mean results of the electrical signals measured before (185 mm from the left edge) and after (315 mm from the left edge) a symmetrical notch, respectively when the S_0 mode is launched for a plate containing a 2.5 mm deep notch (p=1/6).

In figure 8a the first and the second wave packets correspond to the incident (in) and the first reflection (re_1) from the left notch edge of the S_0 mode. Furthermore, multiple reflections from the notch and plate edges are observed. Figure 8b display the second transmission (tr_2) after the notch of the S_0 mode. Here again, no mode conversion is observed and the addition mean results are not shown for brevity and conciseness.

Since each wave packet is identified and quantified, the Hilbert transform is used to evaluate the wave packets' peaks. These measurements are used to compute the power flux energy which are shown hereafter.

http://dx.doi.org/10.5772/54582

4.1.3. Asymmetrical notches

Figures 9a, b, c and d display the mean results of addition (Figs. 9a and d) and subtraction (Figs. 9b and c) of the electric signals measured before the notch when the A_0 and S_0 modes are launched, respectively and encounter a 3 mm deep asymmetric notch (p=3/6). Its shown that the application of the addition and subtraction operations allow to separate the anti-symmetrical contribution from the symmetrical one. On these figures, the incident (in), the first reflection (re_1) of the non-converted and the converted modes are identified and quantified. Furthermore, multiple reflections from the edges of the notch and the plate are also observed.

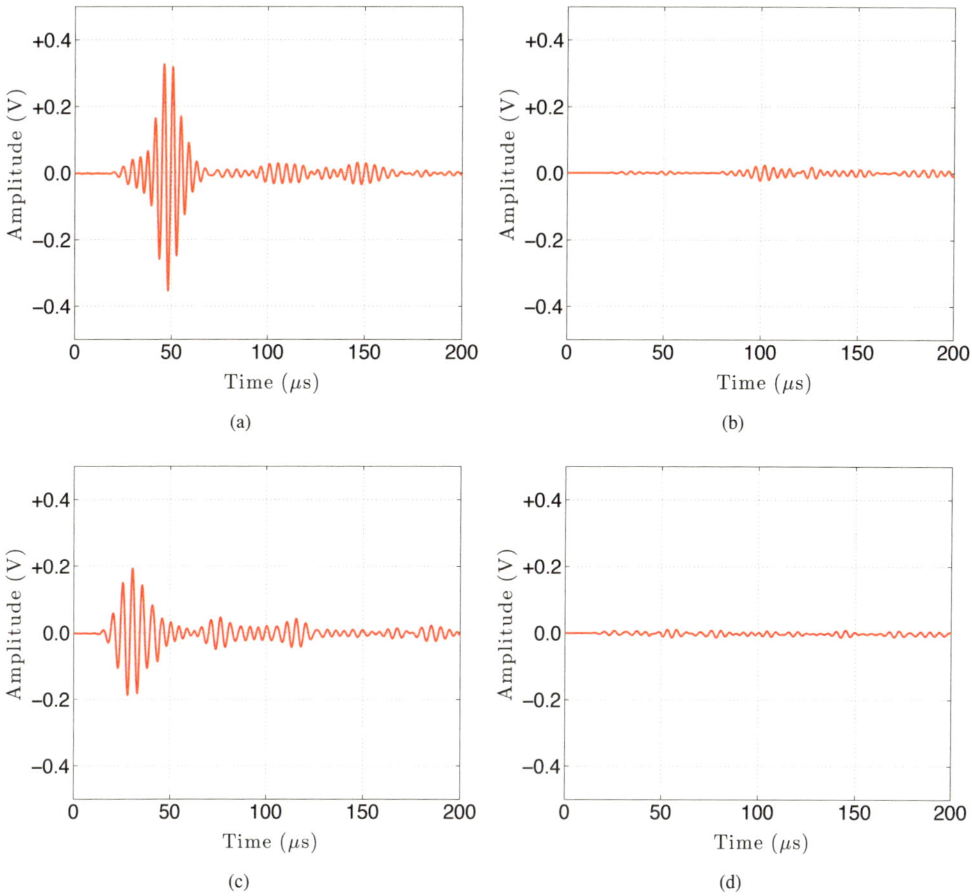

Figure 9. Mean results before the notch of 3 mm deep at a distance of 185 mm from the left of the plate edge: (a) addition and (b) subtraction when the A_0 mode is lunched, (c) subtraction and (d) addition when the S_0 mode is lunched.

4.2. Determination of the power coefficients

The determination of the power coefficients from only the temporal normal or tangential displacement at a given location on the plate surface was presented in previous works [13–15]. Hence, the power reflection and transmission coefficients of the nth Lamb mode when the nth mode is driven for the symmetric or asymmetric step-down damage are:

$$R_{nn}^{Nor} = \left| \frac{\widetilde{W}_{n(re)}(x_1, d)}{\widetilde{W}_{n(in)}(x_1, d)} \right|^2 , \tag{1}$$

and

$$T_{nn}^{Nor} = \left| \frac{\widetilde{W}_{n(tr)}(x_1, dp)}{\widetilde{W}_{n(in)}(x_1, d)} \right|^2 \left| \frac{\widetilde{W}^0_{n(in)}(d)}{\widetilde{W}^0_{n(tr)}(dp)} \right|^2 \frac{\widetilde{P}_{nn}|_{2dp}}{\widetilde{P}_{nn}|_{2d}} , \tag{2}$$

where the first subscript (n) of R and T designates the incident Lamb mode and the second one (n) designates the reflected or the transmitted mode. $\widetilde{W}^0_n$ corresponds to the modal (eigen value) normal harmonic displacement and $\widetilde{W}_n$ corresponds to the normal displacement. The tilde symbol ($\widetilde{\ }$) indicates that the corresponding quantity is calculated at the central frequency. The superscript Nor indicates that the normal displacement is used to compute the above coefficients. in, re and tr denote the terms incident, reflected and transmitted, respectively. x_1 is the propagation direction. P_{nn} is the average power flow of the nth Lamb mode at the central frequency [29]. The notations $|_{2d}$ and $|_{2dp}$ indicate that the average power flow is calculated for a plate thickness of $2d$ and $2dp$, respectively.

Moreover, the asymmetrical discontinuities produce a converted fundamental Lamb mode noted m. Therefore, the power reflection and transmission coefficients of the mth Lamb mode when the nth mode is driven for an asymmetric step-down damage are computed as:

$$R_{nm}^{Nor} = \left| \frac{\widetilde{W}_{m(re)}(x_1, d)}{\widetilde{W}_{n(in)}(x_1, d)} \right|^2 \left| \frac{\widetilde{W}^0_{n(in)}(d)}{\widetilde{W}^0_{m(re)}(d)} \right|^2 \frac{\widetilde{P}_{mm}|_{2d}}{\widetilde{P}_{nn}|_{2d}} , \tag{3}$$

and

$$T_{nm}^{Nor} = \left| \frac{\widetilde{W}_{m(tr)}(x_1, dp)}{\widetilde{W}_{n(in)}(x_1, d)} \right|^2 \left| \frac{\widetilde{W}^0_{n(in)}(d)}{\widetilde{W}^0_{m(tr)}(dp)} \right|^2 \frac{\widetilde{P}_{mm}|_{2dp}}{\widetilde{P}_{nn}|_{2d}} . \tag{4}$$

Instead of the normal displacement, the tangential one could be also used to compute these coefficients. Then, the corresponding values are noted R_{nn}^{Tan}, T_{nn}^{Tan}, R_{nm}^{Tan} and T_{nm}^{Tan}.

In fact, $\widetilde{W}^0_n$ and $\widetilde{W}^0_m$ are determined analytically while, $\widetilde{W}_n$ and $\widetilde{W}_m$ are measured experimentally or computed numerically. When studies are carried out in transient regime, the measured or computed displacements are taken at the central working frequency, which corresponds to the maximum of the Lamb wave packet envelope.

4.3. Numerical computations

4.3.1. General description of the simulation

A lossless aluminium plate is considered with thickness ($2d$) and length ($2L$) equal to 6 mm and 500 mm, respectively. The longitudinal velocity (c_L), the transverse velocity (c_T) and the density (ρ) of this plate are equal to 6422 m/s, 3110 m/s and 2695 kg/m^3, respectively. Figs. 10a and 12a illustrate a plate containing a meshed symmetrical and asymmetrical notches, respectively. Here, notches have a variable width equal to w. The meshing of the symmetrical and asymmetrical steps down are illustrated on Figs. 10b and 12b, respectively. The meshing of the symmetrical and asymmetrical steps up are illustrated on Figs. 10c and 12c, respectively. In all steps, the plate thickness changes abruptly either from $2d$ to $2dp$ or from $2dp$ to $2d$. d is the half thickness of the plate and p takes values from 0 to 1 with a constant increment. In this work, more than 10 isoparametric eight-node quadrilateral elements are used for one wavelength to maintain a large accuracy of the Finite Element (FE) results.

In this paper, the fundamental Lamb mode, A_0 or S_0, is launched from the edge of the plate by the application of the appropriate displacement shapes. The excitation signals (e_n) are windowed by a Hanning temporal function and the central working frequency is f_c=200 kHz. The tone burst number is N_{cyc}=10 and the time sampling period is Δt=0.1 μs.

4.3.2. Symmetrical notches

Figure 11 shows a comparison between the reflection and transmission coefficients for the symmetrical step down and step up. These coefficients are obtained by launching either the A_0 mode (Fig. 11a) or the S_0 mode (Fig. 11b). From the results of Fig. 11, relations of equality can be observed and expressed by the following equations:

$$R_n^D = R_n^I \,, \tag{5}$$

$$T_n^D = T_n^I \,, \tag{6}$$

where D and I designate Direct and Inverse for step down and step up damage, respectively.

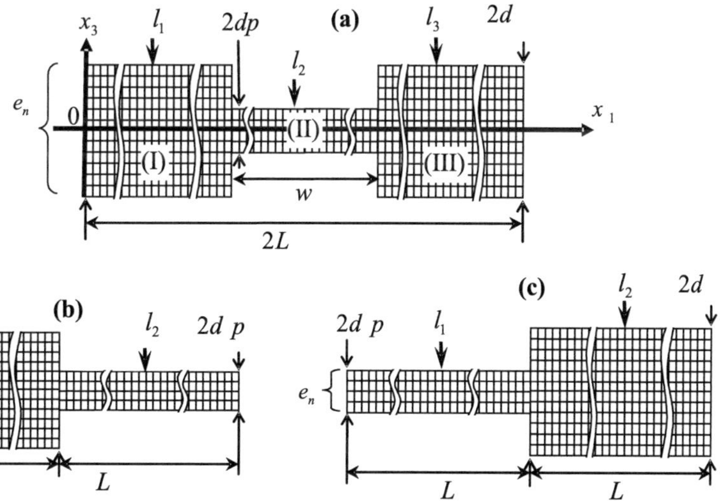

Figure 10. Mesh of an aluminium plate with a symmetrical notch (a), a symmetrical step down (b) and a symmetrical step up (c).

As said in the introduction, the above equations are valid if the frequency of the selected fundamental Lamb modes are taken under the cut-off frequencies of A_1 and S_1 modes. The observed small errors on Figs. 11a and b are due to the mesh precision of the FE meshing at the left edge of the plate. Indeed, the applied excitation in the symmetrical step up case is performed on fewer nodes along the thickness than for the symmetrical step down case.

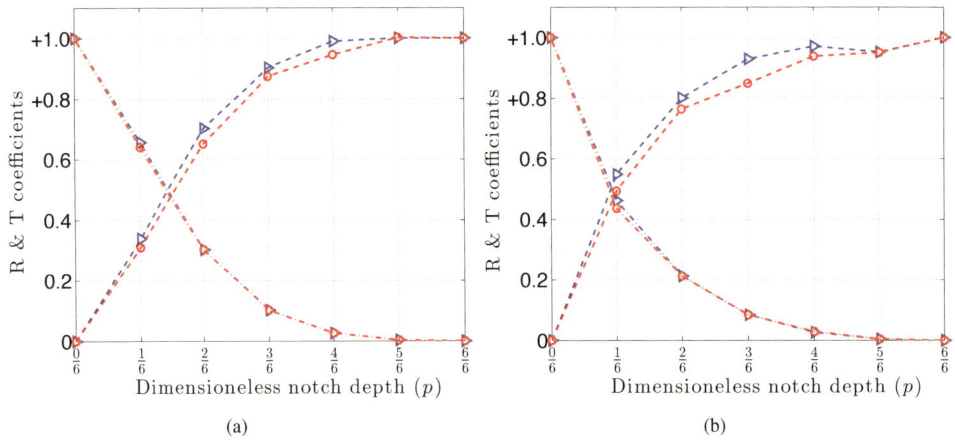

Figure 11. Comparison between the reflection (—·) and transmission (——) power coefficients computed at 200 kHz for a symmetric step down (▷) and step up (o) when the A_0 (a) and S_0 (b) modes are launched.

The symmetrical notch is now decomposed as the superposition of a symmetrical step down and step up as shown on Fig. 10. The incident power I performed on the symmetrical step down is equal to unity, while the incident power transmitted to the symmetrical step up becomes equal to T_1. Using these remarks and the equality relation between the symmetrical step down and step up (Eqs. (5) and (6)), it is easy to obtain the following relationship:

$$\frac{R_2}{R_1} = \frac{T_1}{I} = T_1 \,. \tag{7}$$

Indeed, this relation shows that the reflected power compared to the incident power is equivalent for the symmetrical step down and step up. Furthermore, using the power balance at the interface of the regions (I) and (II), i.e. $R_1 + T_1 = 1$, and at the interface of the regions (II) and (III), i.e. $R_2 + T_2 = T_1$, R_2 and T_2 can be simply expressed as: $R_2 = R_1(1 - R_1)$ and $T_2 = (1 - R_1)^2$. Consequently, these relations show that with one FE simulation of the symmetrical step down or the symmetrical step up, all coefficients (R_1, R_2, T_1 and T_2) of the symmetrical notch can be derived.

4.3.3. Asymmetrical notches

Figures 13 and 14 show a comparison between the power coefficients for the asymmetrical step down and step up when the driven mode is the A_0 and the S_0 mode, respectively. Figures. 13a and 14a illustrate the reflection coefficients and Figures. 13b and 14b depict the transmission coefficients. On Figs. 13, a significant reflection of the driven mode A_0 and a low mode conversion from the A_0 to the S_0 mode are observed for the asymmetrical step up. On the contrary, a significant mode conversion

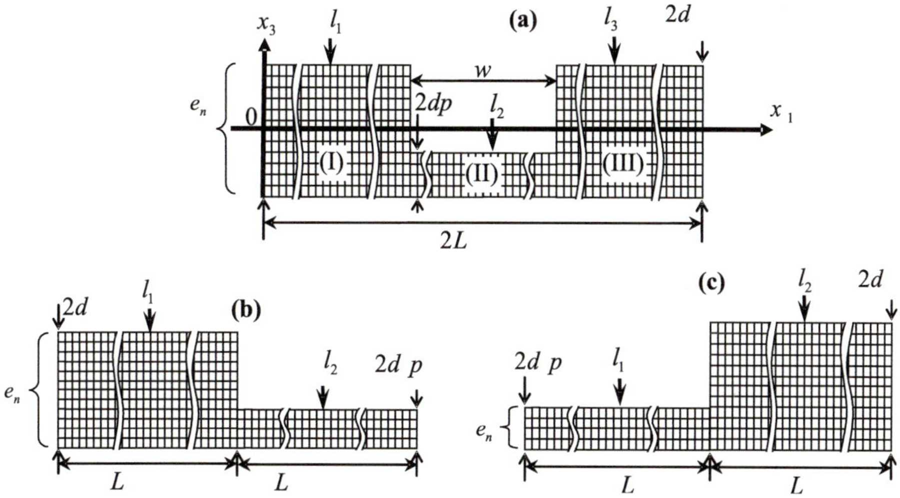

Figure 12. Mesh of an aluminium plate with an asymmetrical notch (a), an asymmetrical step down (b) and an asymmetrical step up (c).

from the S_0 to the A_0 mode for the transmission region is shown on Figs. 14. Therefore, the A_0 mode is more sensitive to the asymmetrical step up than the S_0 mode on the basis of the power reflection. However, the S_0 mode is more sensitive to the asymmetrical step up than the A_0 mode on the basis of the power conversion. The behaviour of the A_0 and S_0 modes towards the asymmetrical step down and step up seems to be inverted.

It is worth mentioning that a simple relation can be observed and expressed when the nth Lamb mode is driven as $R^D_{nm} \cong R^I_{nm}$ and $T^D_{nn} \cong T^I_{nn}$. The observed small errors on Figs. 13 and 14 are due to the mesh precision of the FE meshing at the left edge of the plate as mentioned in the symmetrical case.

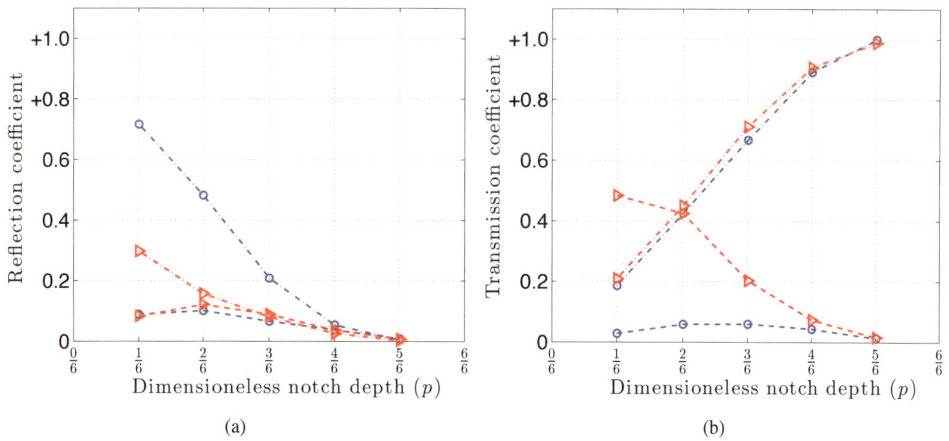

Figure 13. Comparison between the power reflection (a) and the transmission (b) coefficients of the A_0 (—·) and S_0 (— —) modes computed for an asymmetric step down (▷) and step up (○) when the A_0 mode is launched.

In this section, the asymmetrical notch is studied as a superposition of the asymmetrical step down and step up as shown on Fig. 12. The incident power I_n performed on the asymmetrical step down is equal

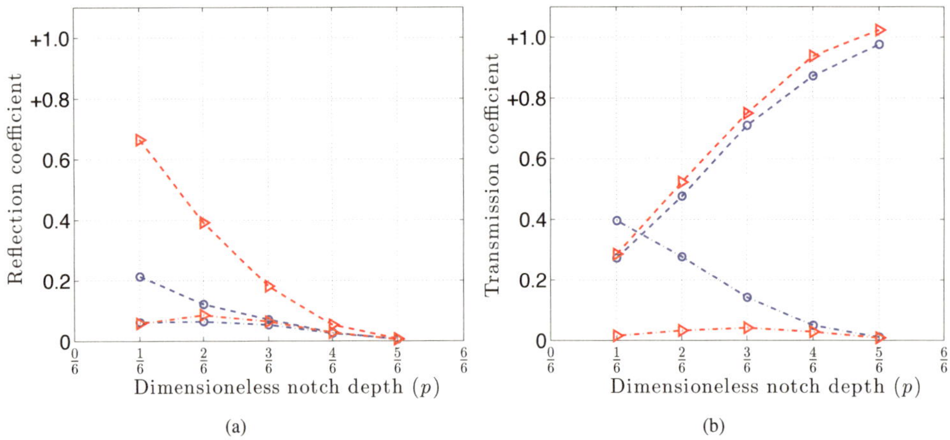

Figure 14. Comparison between the power reflection (a) and the transmission (b) coefficients of the A_0 (—·) and S_0 (——) modes computed for an asymmetric step down (▷) and step up (○) when the S_0 mode is launched.

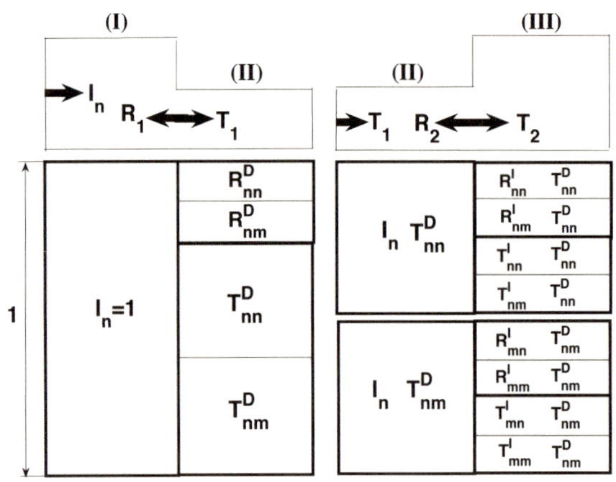

Figure 15. Schematic power balance of the combination of an asymmetric step down and step up.

to unity, while the incident power for the asymmetrical step up becomes equal to $T^D_{nn} + T^D_{nm}$ which corresponds to $T_{1,nn} + T_{1,nm}$ (Fig. 15). Hence, each transmitted wave packet produces four new wave packets. Therefore, the power balance can be expressed for an asymmetrical step up as:

$$(R^I_{nn} + R^I_{nm} + T^I_{nn} + T^I_{nm})T^D_{nn} = T^D_{nn}\,, \tag{8}$$

$$(R^I_{mn} + R^I_{mm} + T^I_{mn} + T^I_{mm})T^D_{nm} = T^D_{nm}\,, \tag{9}$$

Then, the power balance of a constructed asymmetrical notch is derived [14] as:

$$R^D_{nn} + R^D_{nm} + (R^I_{nn} + R^I_{nm} + T^I_{nn} + T^I_{nm})T^D_{nn} + (R^I_{mn} + R^I_{mm} + T^I_{mn} + T^I_{mm})T^D_{nm} = 1. \tag{10}$$

Indeed, all the coefficients of the asymmetrical notch can be derived by analogy from the asymmetrical step down and step up results when the fundamental Lamb mode A_0 or S_0 is launched. Hence, all the coefficients of the complete asymmetrical notch can be computed as for example: $R_{2,mn} = R^I_{mn}\,T^D_{nm}$, $T_{2,mn} = T^I_{mn}\,T^D_{nm}$, $R_{2,mm} = R^I_{mm}\,T^D_{nm}$ and $T_{2,mm} = T^I_{mm}\,T^D_{nm}$.

Furthermore, all the multiple reflections and transmissions and mode conversions of the FEM can now be identified and quantified. This construction method can be generalised to study the interaction of the fundamental Lamb modes with several discontinuities only by using the asymmetrical step down and step up results.

However, the construction technique is valuable only for the fundamental Lamb modes taken under their cut-off frequencies and the notch width verifying:

$$w > w_{lim} = \frac{1}{4} N_{cyc}\,\mathtt{min}(\lambda_{A_0}, \lambda_{S_0})\,, \tag{11}$$

where $\mathtt{min}$ designates minimum. If the rectangle window is used instead of the Hanning one, this limit is given by $\frac{1}{2} N_{cyc}\,\lambda$. In the case of the symmetrical notch, the notch width must verify $w > w_{lim} = \frac{1}{4} N_{cyc}\,\lambda_{(A_0\ \mathtt{or}\ S_0)}$ when the A_0 or S_0 mode is launched.

4.4. Comparisons and validation

Figures. 16a and b show power coefficient comparisons between numerical and experimental results of the interaction of the A_0 and S_0 modes, respectively with different depths of symmetrical notches. Several experimental measurements (20) are performed to determine the displayed error bars. Good agreement is found for both the A_0 and S_0 cases.

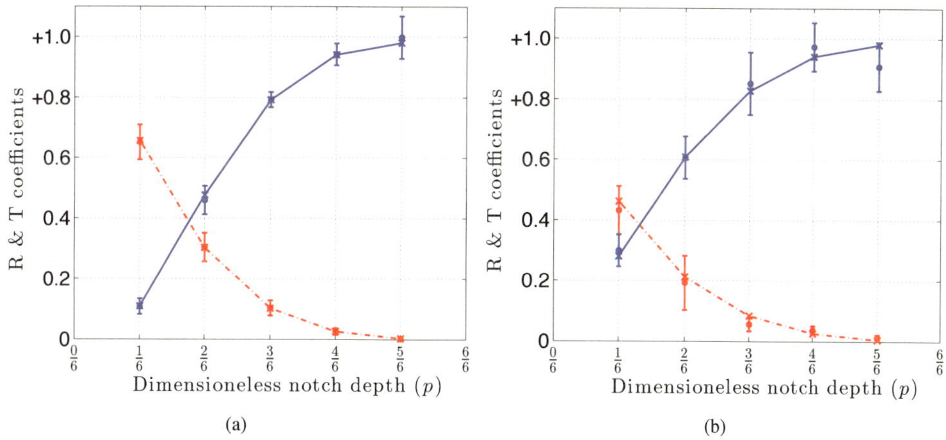

Figure 16. Comparison between numerical and experimental results of the power reflection (—·) and transmission (—) coefficients of the modes A_0 (a) and S_0 (b) when they are launched and interact with symmetrical notches.

Figures 17a and b show power reflection coefficient comparisons between numerical and experimental results of the interaction of the A_0 and S_0 modes, respectively with different depths of asymmetrical notches. Here again, good agreement is found for both the A_0 and S_0 driven modes.

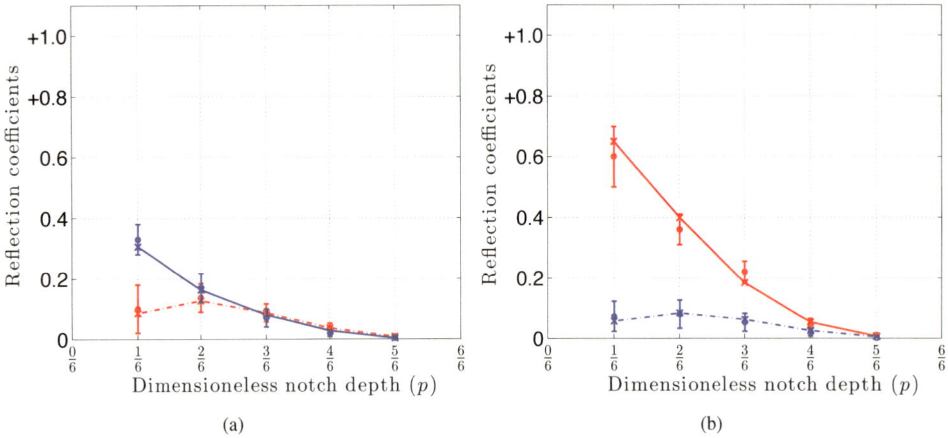

Figure 17. Comparison between numerical and experimental results of the power reflection coefficients of the A_0 (—) and S_0 (— —) modes when the A_0 (a) or S_0 (b) mode is launched and interacts with asymmetrical notches.

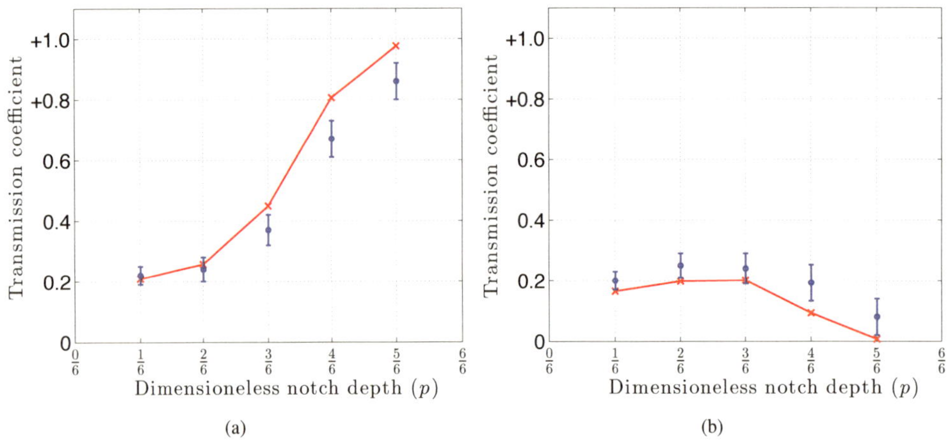

Figure 18. Comparison between numerical (—) and experimental (•) results of the power transmission coefficients of the A_0 (a) and S_0 (b) modes when the A_0 mode is launched and interacts with asymmetrical notches.

The numerical results [14] have shown that the study of the transmission region is very difficult and requires a careful analysis. The power transmission coefficients when the A_0 mode is launched are depicted on Figures 18a and b and show an agreement with the numerical results. However, when the S_0 mode is launched, errors due to the experimental conditions do not permit the obtaining of a sufficient precision to separate and quantify the existent Lamb modes in this region. In fact, the high velocity of the S_0 mode increases the interferences between reflections from the start and the end of the notch edges.

5. Conclusion

Transducers piezoceramic based emitter is developed to excite one selected Lamb mode. To this end, two piezoelectric transducers are cut and placed on the opposite sides of the tested plate. Then the in-phased and in-anti-phased electrical signals are applied to select the A_0 or the S_0 mode. Noteworthy that a calibration process is carried to avoid the receiver effects for the reflection study. The power refection and transmission coefficients are then obtained by using either experimental measurements or numerical predictions and normal mode expansions. This has the advantage to allow a direct comparison between numerical and experimental results. Symmetrical and asymmetrical discontinuities are both investigated. The construction of power coefficients of symmetrical notches from those obtained for one elementary configuration is carried out successfully. Furthermore, the power coefficients for asymmetrical notches are constructed from those obtained for two elementary configurations (step down and step up). On the contrary of the symmetrical discontinuities, the asymmetrical ones enable mode conversions. The experimental measurements are confronted with success to numerical predictions.

Author details

Farouk Benmeddour[1,⋆], Emmanuel Moulin[1],
Jamal Assaad[1] and Sonia Djili[2]

1 IEMN, OAE Department, CNRS UMR 8520, University of Valenciennes and Hainaut Cambrésis, Le Mont Houy 59313 Valenciennes Cedex 9, France
2 Scientific and Technical Center of Research on Welding and Control, BP 64, route de Daly Brahim. Chéraga, 16000 Algies, Algeria

References

[1] Lamb H. On waves in an elastic plate. Proc Roy Soc London. 1917;A 93:114–128.

[2] Diligent O, Grahn T, Boström A, Cawley P, Lowe MJS. The low-frequency reflection and scattering of the S0 Lamb mode from a circular through-thickness hole in a plate: Finite Element, analytical and experimental studies. J Acoust Soc America. 2002;112 (6):2589–2601.

[3] Hayashi T, Kawashima K. Single mode extraction from multiple modes of Lamb wave and its application to defect detection. JSME Int J. 2003;46 (4):620–626.

[4] Castaings M, Le Clézio E, Hosten B. Modal decomposition method for modeling the interaction of Lamb waves with cracks. J Acoust Soc America. 2002;112 (6):2567–2582.

[5] Wang L, Shen J. Scattering of elastic waves by a crack in a isotropic plate. Ultrasonics. 1997;35:451–457.

[6] Cho Y, Rose JL. An elastodynamic hybrid boundary element study for elastic wave interactions with a surface breaking defect. International Journal of Solids and Structures. 2000;37:4103–4124.

[7] Hansch MKT, Rajana KM, Rose JL. Characterization of aircraft joints using ultrasonic guided waves and physically bases feature extraction. IEEE Ultrasonics Symposium. 1994;p. 1193–1196.

[8] Cho Y. Estimation of Ultrasonic guided wave mode conversion in a plate with thickness variation. IEEE Trans Ultrason Ferroelectr Freq Control. 2000;17 (3):591–603.

[9] Leduc D, Hladky AC, Morvan B, Izbicki JL, Pareige P. Propagation of Lamb waves in a plate with periodic grating: Interpretation by phonon. J Acoust Soc America. 2005;118 (4):2234–2239.

[10] Alleyne DN, Cawley P. A 2-dimensional Fourier transform method for the quantitative measurement of Lamb modes. IEEE Ultrasonics Symposium. 1990;p. 1143–1146.

[11] Lowe MJS, Diligent O. Low-frequency reflection characteristics of the S0 Lamb wave from a rectangular notch in a plate. J Acoust Soc America. 2002;111(1):64–74.

[12] Jin J, Quek ST, Wang Q. Wave boundary element to study Lamb wave propagation in plates. J Sound and Vibration. 2005;288:195–213.

[13] Benmeddour F, Grondel S, Assaad J, Moulin E. Study of the fundamental Lamb modes interaction with symmetrical notches. NDT & E International. 2008;41(1):1–9.

[14] Benmeddour F, Grondel S, Assaad J, Moulin E. Study of the fundamental Lamb modes interaction with asymmetrical discontinuities. NDT & E International. 2008;41(5):330–340.

[15] Benmeddour F, Grondel S, Assaad J, Moulin E. Experimental study of the A0 and S0 Lamb waves interaction with symmetrical notches. Ultrasonics. 2009;49(2):202 – 205.

[16] El Youbi F, Grondel S, Assaad J. Signal processing for damage detection using two different array transducers. Ultrasonics. 2004;42:803–806.

[17] Si-Chaib MO, Djelouah H, Bocquet M. Applications of ultrasonic reflection mode conversion transducers in NDE. NDT & E International. 2000;33(2):91 – 99.

[18] Duquennoy M, Ouaftouh M, Qian ML, Jenot F, Ourak M. Ultrasonic characterization of residual stresses in steel rods using a laser line source and piezoelectric transducers. NDT & E International. 2001;34(5):355 – 362.

[19] Yan T, Theobald P, Jones BE. A self-calibrating piezoelectric transducer with integral sensor for in situ energy calibration of acoustic emission. NDT & E International. 2002;35(7):459 – 464.

[20] Blomme E, Bulcaen D, Declercq F. Air-coupled ultrasonic NDE: experiments in the frequency range 750kH-2MHz. NDT & E International. 2002;35(7):417 – 426.

[21] Martínez O, Akhnak M, Ullate LG, de Espinosa FM. A small 2D ultrasonic array for NDT applications. NDT & E International. 2003;36(1):57 – 63.

[22] Bhalla S, Soh CK. High frequency piezoelectric signatures for diagnosis of seismic/blast induced structural damages. NDT & E International. 2004;37(1):23 – 33.

[23] Sun M, Staszewski WJ, Swamy RN, Li Z. Application of low-profile piezoceramic transducers for health monitoring of concrete structures. NDT & E International. 2008;41(8):589 – 595.

[24] Rathod VT, Mahapatra DR. Ultrasonic Lamb wave based monitoring of corrosion type of damage in plate using a circular array of piezoelectric transducers. NDT & E International. 2011;44(7):628 – 636.

[25] Chung CH, Lee YC. Fabrication of poly(vinylidene fluoride-trifluoroethylene) ultrasound focusing transducers and measurements of elastic constants of thin plates. NDT & E International. 2010;43(2):96 – 105.

[26] Ribichini R, Cegla F, Nagy PB, Cawley P. Experimental and numerical evaluation of electromagnetic acoustic transducer performance on steel materials. NDT & E International. 2012;45(1):32 – 38.

[27] Lee YC, Lin Z. Miniature piezoelectric conical transducer: Fabrication, evaluation and application. Ultrasonics. 2006;44, Supplement:e693 – e697. Proceedings of Ultrasonics International (UI) and World Congress on Ultrasonics (WCU).

[28] Bowen CR, Bradley LR, Almond DP, Wilcox PD. Flexible piezoelectric transducer for ultrasonic inspection of non-planar components. Ultrasonics. 2008;48(5):367 – 375.

[29] Auld BA. Acoustic fields and waves in solids. vol. II. A Wiley-Interscience publication; 1973.

Chapter 6

Piezoelectric Pressure Sensor Based on Enhanced Thin-Film PZT Diaphragm Containing Nanocrystalline Powders

Vahid Mohammadi, Saeideh Mohammadi and Fereshteh Barghi

Additional information is available at the end of the chapter

http://dx.doi.org/10.5772/54755

1. Introduction

During recent years, the study of micro-electromechanical systems (MEMS) has shown that there are significant opportunities for micro sensors and microactuators based on various physical mechanisms such as piezoresistive, capacitive, piezoelectric, magnetic, and electrostatic. In addition, specialized processes, novel materials, and customized packaging methods are routinely used. MEMS themes include miniaturization, multiplicity, and microelectronic manufacturing and integration. Huge technology opportunities for MEMS are present in automotive applications, medicine, defense, controls, and communications. Other applications include biomedical pressure sensors and projection displays. For several excellent MEMS overviews of both core technologies and emerging applications, the readers are encouraged to consult references [1–4].

MEMS can be classified in two major categories: sensors and actuators. MEMS sensors, or microsensors, usually rely on integrated microfabrication methods to realize mechanical structures that predictably deform or respond to a specific physical or chemical variable. Such responses can be observed through a variety of physical detection methods including electronic and optical effects. Structures and devices are designed to be sensitive to changes in resistance (piezoresistivity), changes in capacitance, and changes in charge (piezoelectricity), with an amplitude usually proportional to the magnitude of the stimulus sensed. Examples of microsensors include accelerometers, pressure sensors, strain gauges, flow sensors, thermal sensors, chemical sensors, and biosensors. MEMS actuators, or microactuators, are usually based on electrostatic, piezoelectric, magnetic, thermal, and pneumatic forces. Examples of microactuators include positioners, valves, pumps, deformable mirrors, switches, shutters, and resonators.

Among all MEMS technologies, the Piezoelectric MEMS offer more advantages than others [5]. New materials and new processing technologies continue to enlarge the offer of highly performing piezoelectrics. In parallel, the range of applications is growing and there is an increasing need for functioning under varied conditions and wider operation ranges, or sometimes in the extreme environments with high temperatures, high frequencies, and high electric fields or pressures. One of the recent developments is the use of piezoelectric thin films for microsensors, and microsystems. They have attracted a great interest for microsystems thanks to their reversible effect. Zinck et al. [6] presented the fabrication and characterization of silicon membranes actuated by piezoelectric thin-films. Combining piezoelectric thin-films with micromachined silicon membranes has resulted in novel micro-devices such as motors, accelerometers, pressure sensors, micro pumps, actuators and acoustic resonators.

The MEMS pressure sensors have been developed in the 1970's. Many works have been presented on developing these devices. For instance, Ravariu et al. [7] modeled a pressure sensor for computing the blood pressure by using of ANSYS simulation in order to estimate the mechanical stress in their structure. Liu et al. [5] designed two novel piezoelectric microcantilevers with two piezoelectric elements (bimorph or two segments of Lead Zirconate Titanate (PZT) films) and three electric electrodes. The PZT thin films are very attractive due to their larger piezoelectric properties compared to the most conventional piezoelectric materials.

This chapter is concerned on the application of the PZT and/or nanocrystalline-powders-enhanced (ncpe-) PZT thin-films in micro-electro-mechanical sensors such as multilayer diaphragm pressure sensors. This is addressed to the users, designers, researchers, developers and producers of piezoelectric materials who are interested to work on PZT-based devices and systems in various applications.

2. Piezoelectric ceramic materials

Piezoelectric materials have been integrated with silicon microelectromechanical systems (MEMS) in both microsensor and microactuator applications [8]. An understanding of the development of crystal structure, microstructure, and properties of these films is necessary for the MEMS structural design and process integration. In this section, piezoelectric thin films are reviewed, beginning with a short discussion of piezoelectric materials in general, followed by thin-film processing, structure and property development (focusing on solution deposition methods) and lastly, piezoelectric properties of thin films.

2.1. Piezoelectric materials

A piezoelectric is a material that develops a dielectric displacement (or polarization) in response to an applied stress and, conversely, develops a strain in response to an electric field [9]. To achieve the piezoelectric response, a material must have a crystal structure that lacks a center of symmetry. Twenty of the possible 32 point groups that describe a crystal's symmetry fulfill this requirement and are piezoelectric [10]. The importance of the crystal structure to piezoelectricity extends into understanding the constitutive equations describing the piezoelectric's response (Eqs. 1 and 2). For example, the application of an electric field along a certain crystallographic direction may cause a strain in more than one direction. Such relationships between applied electric field and strain, and between applied stress and dielectric displacement (or polarization), are specific to the piezoelectric's crystal structure, and the magnitude of the response is given by a material's piezoelectric coefficients (dij). The piezoelectric constitutive relationships are described in detail in several publications [9–11].

$$S_i = S_{ij}^{E} T_j + d_{ki} E_k \quad (1)$$

$$D_1 = d_{lm} T_m + \varepsilon_{ln}^{T} E_n \quad (2)$$

where i, j, m = 1, ..., 6 and k, l, n = 1, 2, 3. Here, S, D, E, and T are the strain, dielectric displacement, electric field, and stress, respectively, and S_{ij}^{E}, d_{ki} and ε_{ln}^{T} are the elastic compliances (at constant field), the piezoelectric constants, and dielectric permittivities (at constant stress), respectively.

A wide variety of materials are piezoelectric, including poled polycrystalline ceramics (e.g. lead zirconate titanate, PZT), single-crystal or highly oriented polycrystalline ceramics (e.g. zinc oxide and quartz), organic crystals (e.g. ammonium dihydrogen phosphate), and polymers (e.g. polyvinylidiene fluoride), as shown in Table 1 [12].

Material	Formula	Form	Piezoelectric constant (pm/V or pC/N)
Ammonium dihydrogen phosphate (ADP)	$NH_4H_2PO_4$	Single crystal	$d_{36} = 48$
Barium titanate	$BaTiO_3$	Single crystal	$d_{15} = 587$
Barium titanate	$BaTiO_3$	Polycrystalline ceramic	$d_{15} = 270$
Lead zirconate titanate (PZT)	$PbZr_{0.6}Ti_{0.40}O_3$	Polycrystalline ceramic	$d_{33} = 117$
Lead lanthanum zirconate titanate (PLZT)	$Pb_{0.925}La_{0.5}Zr_{0.56}Ti_{0.44}O_3$	Polycrystalline ceramic	$d_{33} = 545$
Polyvinylidene fluoride	$(CH_2CF_2)_n$	Oriented film	$d_{31} = 28$
Potassium dihydrogen phosphate (KDP)	KH_2PO_4	Single crystal	$d_{36} = 21$
Quartz	SiO_2	Single crystal	$d_{11} = 2.3$
Zinc oxide	ZnO	Single crystal	$d_{33} = 12$

Table 1. Properties of some piezoelectric materials (from Ref. [12])

In general, these piezoelectrics belong to one of two categories: those that are also ferroelectric and those that are not. Ferroelectric materials have the further restrictions that their crystal structures have a direction of spontaneous polarization (10 of the point groups are polar) and that their polarization can be oriented by the application of an electric field and will remain oriented to some degree when that field is removed [13]. This property of polarization reversal and remanence cannot be predicted by the material's structure; it must be determined experimentally. The polarization-field hysteresis loop illustrated and described in Figure 1 is the practical demonstration of ferroelectricity.

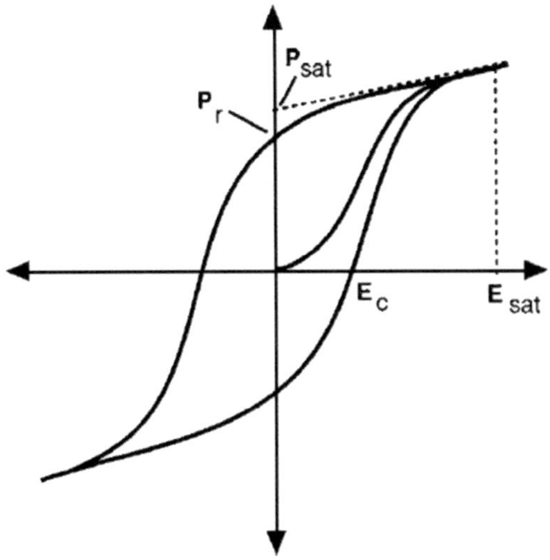

Figure 1. A polarization-electric field hysteresis loop. When a field is applied to a randomly oriented polycrystalline material, domains (regions of uniform polarization) align with respect to the applied field to give a net polarization that saturates at value P_{sat}. When the field is reduced back to zero, a remanent polarization (P_r) persists; when the field is applied in the opposite sense, the polarization reduces to zero with the application of the coercive field (E_c) and then switches directions and saturates.

What importance is the distinction between ferroelectric and non-ferroelectric for piezoelectric materials? The ferroelectric's ability to orient its polarization allows it to be poled (by application of an electric field typically at elevated temperature) so that the polar axes in a random polycrystalline material can be oriented and produce a net piezoelectric response.

The application of piezoelectrics in MEMS requires that the material be processed within the constraints of microfabrication and have the properties necessary to produce a MEMS device with the desired performance. Microfabrication nominally requires that a thin film be prepared with conducting electrodes and that the film be ferroelectric or oriented (textured) properly for the desired piezoelectric response. Fabrication of some devices requires that the film be patterned and that it withstand processes such as encapsulation and wire bonding. Zinc oxide with a preferred orientation fits this first requirement and has been used as piezoelectric film for many years [14] and frequently in MEMS [15]. A second consideration is the properties that include the piezoelectric constants, as well as the dielectric properties and elastic properties.

The specific property requirements depend on the device, but in general large piezoelectric constants are desired for piezoelectric MEMS. Ferroelectric ceramics, particularly those with the perovskite (ABO_3) structure, are known to have very high piezoelectric constants. The ferroelectric ceramics receiving the most widespread use as bulk piezoelectrics as well as thin-film piezoelectrics are in the lead zirconate titanate ($PbZr_{1-x}Ti_xO_3$, PZT) system.

The PZT family of ceramics is widely used due to its excellent piezoelectric and dielectric properties [9]. PZT materials have the perovskite structure (ABO_3) in cubic, tetragonal, rhombohedral, and orthorhombic forms, depending on the temperature and composition, as shown in the Figure 2.

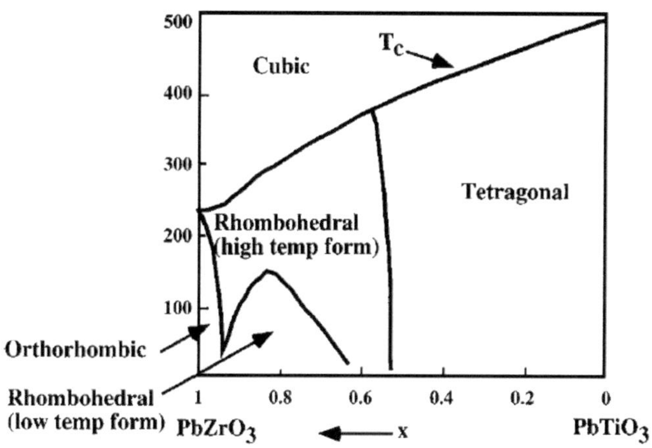

Figure 2. Phase diagram for the PbZrO3-PbTiO3 system (from Ref. [9]). The nearly vertical phase boundary between the rhombohedral and tetragonal phases is called the morphotropic phase boundary.

Extensive research has been carried out to determine the effects of composition (Zr/Ti) and small amounts of additives on the electrical and mechanical properties [9, 16-18]. Several important points should be noted. Compositions near the morphotropic phase boundary (i.e. the boundary between rhombohedral and tetragonal phases at $PbZr_{0.53}Ti_{0.47}O_3$) have the largest piezoelectric constants and dielectric constants. This enhancement is due to the greater ease of polarization. These compositions are so common that in this chapter PZT is used to refer to compositions close to this boundary.

Several other ferroelectric ceramics have properties that are comparable with those in the PZT system, particularly other perovskite-based ceramics that have morphotropic phase boundaries (MPB). Among those with properties of interest for MEMS are ceramics in the lead magnesium niobate-lead titanate system (MPB at 30 mole% lead titanate) [19], lead zinc niobate-lead titanate system (MPB at 9 mole% lead titanate) [20] and the lead scandium niobate-lead titanate system (MPB at 42 mole% lead titanate) [21]. There are also a number of interesting compositions in the lead lanthanum zirconate titanate system (PLZT) [22]. However, the list of materials of interest for piezoelectric MEMS does not stop here; the literature on piezoelectric MEMS reveals that two materials have dominated: PZT and ZnO. In this part, we focus on the use of PZT in MEMS devices particularly MEMS pressure sensors.

2.2. Thin-film processing, structural evolution, and properties

For piezoelectric MEMS, a key processing challenge is to create a piezoelectric thin film with the desired structure and properties. A revival in research on ferroelectric ceramic thin films began in the early 1980s and has led to significant progress in understanding how to process thin ceramic films and control their electrical properties. This research has been sparked primarily by non-MEMS applications such as FRAMs (ferroelectric random access memories [23]), dynamic RAM (DRAMs [24]) and high dielectric constant decoupling capacitors [25].

The knowledge gained in these pursuits benefits piezoelectric MEMS due to the similarities in the materials used. A series of proceedings volumes on ferroelectric thin films provides a host of information on the topic [26-28]. Here we focus on thin-film processing by solution deposition and on its implication on the structural evolution of the film and resulting properties.

In processing PZT piezoelectric thin films, several methods are available, including physical vapor deposition (PVD), chemical vapor deposition (CVD), and solution deposition (SD). Each method has unique advantages and disadvantages. PVD methods (e.g. sputtering [29]) and CVD methods (e.g. MOCVD [30]) offer uniform thickness films and good step coverage; in addition, these routes are currently standard in microfabrication facilities. However, depositing the correct stoichiometry from the these methods is often challenging. By contrast, SD methods (e.g. sol-gel [31]) offers excellent control of the chemistry of the thin film but is not appropriate when uniform film thickness over surface features is required [32]. A further advantage of SD methods is their simplicity (no vacuum or reactor chambers are required). SD methods have three basic steps: synthesis of a metalorganic solution, deposition onto a substrate by a spin-casting or dip-coating method, and heat-treatment to remove organics and crystallize the ceramic microstructure. The general considerations in processing ceramic coatings by such routes are reviewed elsewhere [33, 34]. In addition, several publications on solution-deposited ferroelectric films can be found in the literature [35–37].

The development of crystal structure, microstructure, and properties is strongly dependent on processing conditions such as the solution chemistry, the thermal treatment and the gas atmosphere, as well as the electrode onto which the film is deposited. The first challenge in structural development is to form the desired perovskite crystal structure and eliminate the metastable pyrochlore (or fluorite [38]) form. On heating, pyrochlore forms at a lower temperature than does perovskite [39–41] and is a common alternative form for many perovskite ferroelectrics, particularly relaxor ferroelectrics. Because this pyrochlore is non-ferroelectric and has a low dielectric constant, both the ferroelectric and dielectric constant are degraded by its presence [40]. For PZT, pyrochlore will transform into perovskite when the film is heated to higher temperatures [39–42]. In many cases, pyrochlore is found preferentially at the surface and goes undetected in X-ray diffraction. Surface pyrochlore may also be indicative of lead oxide evaporation [39, 43]. In nearly all processing schemes, excess lead is added to solutions to accommodate the evaporation and combat pyrochlore formation. This excess also has been shown to enhance the formation of perovskite and improve properties [39], as well as lower the temperature of perovskite crystallization in PZT [44] and $PbTiO_3$ [45].

Thermal treatment conditions also impact the crystalline phase development. Thermal treatments usually consist of at least two steps (usually after drying): one to remove residual bound organics (and, sometimes, solvent) and another to develop the perovskite microstructure. The two-step procedure can be performed on a single layer after deposition or after several layers have been deposited. The first step may not only remove organics but also lead to some pyrochlore crystallization. For PZT, this initial pyrochlore does not prevent complete transformation to perovskite, but in other materials such as lead magnesium niobate, a single heat treatment to high temperature is a better route so that the pyrochlore formation can be minimized [46]. The effect of heating rate, including rapid thermal processing [47, 48], has been explored. An interesting and unexpected effect of thermal treatment on structure was reported by Chen & Chen [49]. They showed that films prepared by an MOD method and deposited on Pt-coated Si form perovskite with a [111] texture when rapidly heated to 600–700º C, whereas a two-step treatment leads to [100] texture.

The electrode or substrate is as important as any processing condition in determining the structure and properties of the film. The electrode materials of choice for integration with Si are platinum (with a thin Ti adhesion layer) [50–53]. Electrodes also potentially impact the crystalline structure through providing nucleation sites and influencing orientation. For example, the close lattice match between (111) Pt and (111) PZT can influence the film texture.

Substrates also play a role. Tuttle et al [54] showed that substrate thermal expansion coefficient influences the stress-state of the film and the resulting crystallographic orientation and switching properties. They found that solution deposited films on Pt-coated MgO were in a state of compression, whereas those deposited on Si-based substrates were in tension.

2.3. Piezoelectric properties and characterization

The piezoelectric response in thin films can be measured by applying a stress to the film and measuring the induced charge (direct effect) or by applying an electric field and measuring the strain induced in the film (converse effect). For PZT thin films, the piezoelectric constants of interest are d_{33} and d_{31}. The first (d_{33}) relates the strain (S_3) in the direction of electric field (E_3) to the electric field strength ($S_3 = d_{33}E_3$) or equivalently relates the induced charge (D_3) on electroded faces perpendicular to an applied stress (T_3) to the stress ($D_3 = d_{33}T_3$). The second (d_{31}) relates the strain (S_1) in the direction perpendicular to the applied field to the field strength ($S_1 = d_{31}E_3$) or relates the induced charge on electrodes parallel to the direction of stress application to the stress ($D_3 = d_{31}T_1$). The piezoelectric effect has been detected in poled and un-poled films. Without poling, a preferred crystallographic orientation, as well as possible alignment during measurement, makes this response possible. Poling requires application of an electric field, typically at higher temperatures, to align domains and develop a net polarization in a polycrystalline film. A considerable amount of research is now underway to try to understand the piezoelectric properties of thin-film ceramic ferroelectrics such as PZT.

For the direct effect, a normal load can be applied onto an electroded piezoelectric film and the charge on the electrodes measured. In this case, the electrical response is parallel to the applied stress and a d_{33} coefficient is determined [55]. For the converse effect, an electric field is applied and a strain in a thin film is measured. For d_{33} determination, laser interferometry methods are used to monitor changes in thickness in a film (on a substrate) upon application of a small ac field [56, 57].

Like other properties, the piezoelectric properties of PZT thin films depend on structural factors. For example, the presence of a non-piezoelectric phase (e.g. pyrochlore) dilutes the piezoelectric response. However, unlike the dielectric and ferroelectric properties of thin films, the measured values for the piezoelectric coefficients are typically lower than those of bulk PZT (see Table 2), and dynamics of domain orientation and switching appear to be more complex in films.

Processing route	Measurement	Poling conditions	Piezoelectric constants (pm/V or pC/N)	Reference
Solution deposition	Direct-normal load	Unpoled ~200 kV/cm for a few min	$d_{33} = 0$ $d_{33} = 400$	72
MOCVD	Direct-normal load	Unpoled ~40 kV/cm for a few min	$d_{33} = 20–40$ $d_{33} = 200$	72
Solution deposition	Direct-flexed substrate	200 kV/cm for 21 h	$d_{31} = -77$	73
Solution deposition	Converse (single beam interferometer, dc bias)	Unpoled	$d_{33} = 80$	75
Solution deposition	Converse (double beam interferometer, dc bias)	Poled 230 kV/cm, 900 s	$d_{33} = 58$	81
RF Magnetron sputtering	Direct (free-standing film beam deflection)	Unpoled (*c*-axis oriented)	$d_{31} = -100$	69
Solution deposition	Converse (single beam interferometer, dc bias)	Unpoled	$d_{33} = 80$	78
Solution deposition	Converse (single beam interferometer, no dc bias)	Unpoled	$d_{33} = 27$	39
Solution deposition	Converse (double beam interferometer, dc bias)	?	$d_{33} = 100$ (0.33 μm thick) $d_{33} = 140$ (7.1 μm thick)	80

Table 2. Reported piezoelectric constants for PZT thin films with compositions near the morphotropic phase boundary

3. Nanocrystalline-powders-enhanced (*ncpe-*) PZT

Piezoceramic materials have attracted much attention for sensing, actuation, structural health monitoring and energy harvesting applications in the past two decades due to their excellent coupling between energy in the mechanical and electrical domains. Among all piezoceramic materials, lead zirconate titanate ($PbZr_{0.52}Ti_{0.48}O_3$, PZT) has been the most broadly studied and implemented, in industrial applications due to its high piezoelectric coupling coefficients. Piezoceramic materials are most often employed as thin films or monolithic wafers. The integration of these thin films on silicon substrates has a great interest to produce piezoelectric microsystems such as membrane base sensors and actuators. Different technologies have been reported to deposit thin PZT films: MOCVD, sol-gel, laser-ablation, sputtering [29-31, 58]. Among these mentioned methods and the numerous different methods, the sol-gel processing technique is the most widely used due to its low densification temperature, the ease at which the film can be applied without costly physical deposition equipment and the capability to fabricate both thin and thick films [59].

In this section, will introduce the nanocrystalline-powders-enhanced (*ncpe-*) PZT. This is the novel technique to enhance the piezoelectric properties of PZT sol-gel derived ceramics through the use of single crystal PZT microcubes as an inclusion in the PZT sol-gel [69]. This novel technique is crucial to enhance the PZT properties in the case of MEMS applications eg. in MEMS pressure sensors. Because the piezoelectric properties of PZT sol-gel derived films are substantially lower than those of bulk materials, which limit the application of sol-gel films. In comparison, single crystal PZT materials have higher piezoelectric coupling coefficients than polycrystalline materials due to their uniform dipole alignment. These nanocrystalline-PZT powders as PZT single crystal cubes are used to enhance the PZT properties.

The nanocrystalline-PZT powders are synthesized through a hydrothermal based method and their geometry and crystal structure is characterized through scanning electron microscopy (SEM) and X-ray diffraction (XRD). A mixture of PZT cubes and sol-gel will then be sintered to crystallize the sol-gel and obtain full density of the ceramic. XRD and SEM analysis of the cross section of the final ceramics will be performed and compared to show the crystal structure and microstructure of the samples. The results will show that the considerable enhancement is achieved with the integration of nanocrystalline-PZT powders to the PZT thin-film. Recently, the amount of more than 200% increase in the d_{33} coupling coefficient is also reported compared to that of pure PZT sol-gel thin-film [70, 71].

3.1. Preparation of PZT solution

The first step is the dissolution of appropriate of lead acetate trihydrate in 2-methoxyethanol solvent at 120°C for 30 minutes. Pb is added at a composition of approximately 10 mol% more than required by stoichiometry to compensate the PbO loss during high temperature annealing. This solution was vacuum distilled until a white paste with enough moisture begins to form. In a separate flask, Zr n-propoxide and Ti isopropoxide were added drop by drop into 2-MOE and stirred at room temperature for 1 hour. While Zr and Ti mixture solution was stirring, acetylacetone as a chelating agent was added to the solution for further stabilizing. Zr/Ti mixture solution is then added to the flask with paste-like Pb lumps. It then refluxed for improved homogeneity and vacuum distilled to eliminate the byproducts and water molecules from reaction. Final solution was filtered using 0.2 μm filter paper to minimize the incorporation of particles and dust during solution preparation.

3.2. Preparation of PZT nanocrystalline powders

In the present study PZT powders derived from the same solution that had been prepared for film deposition. Because compare to the conventional solid state method, fabrication of powder via sol-gel method has the advantages of simple composition control, high reactivity, lower synthesis temperature, high purity, etc. Then, the solution placed in the oven at 120°C overnight. After drying, calcination performed at 650°C for 2 hours.

3.3. Preparation of slurry and film deposition

A sol-gel method combined with PZT powder will be useful for thick film deposition. During the sintering process, atomic diffusion in the PZT powder grain occurs to minimize the surface energy, which promotes crystal bonding at the interface between two adjacent particles. The added sol will increase the driving force of the system due to the presence of nanoscale particles and so lower the required sintering temperature. In addition, the sol will also function as glue, binding the larger particles together and to the substrate. Nanocrystalline PZT powder which has particle size in order of 0.8μm dispersed in sol solution through an attrition mill to reduce the size of powders. Also 1 wt% of a phosphate ester based dispersant was added to get uniform and stable slurry for film deposition. The mass ratio of powder to sol solution fixed at 1:2. In the case of the composite sol-gel route, each microsize PZT grain will act as a site for crystallization.

The film will easily crack if the concentration is higher. However, each layer of the film will become much thinner if the concentration is lower. The resulting solution was finally spin coated onto a substrate at 3000 rpm for 30s, dried at 150°C for 10 min, fired at 380°C for 15 min and annealed at 650°C for 30 min. The spinning/drying/themolysis procedures were repeated until desired thickness achieved. The resulting coating is essentially a 0-3 ceramic/ceramic composite because the sol gel matrix is connected in all 3 directions and the PZT powder is not connected in any 0 directions. Since the sol gel solution and powders are the same materials, the resulting coating will have properties that compare to that of the bulk material. The overall flow chart for the fabrication of PZT 0-3 ceramic/ceramic composite film is shown in Figure 3.

3.3.1. Characterization of ncpe-PZT layer

XRD results are shown in Figure 4. Pattern of powders shows a random orientation, rhombohedral phase of PZT. Film deposited on amorphous glass could not form any crystalline phase. But layers with powders show the main peaks of perovskite structure.

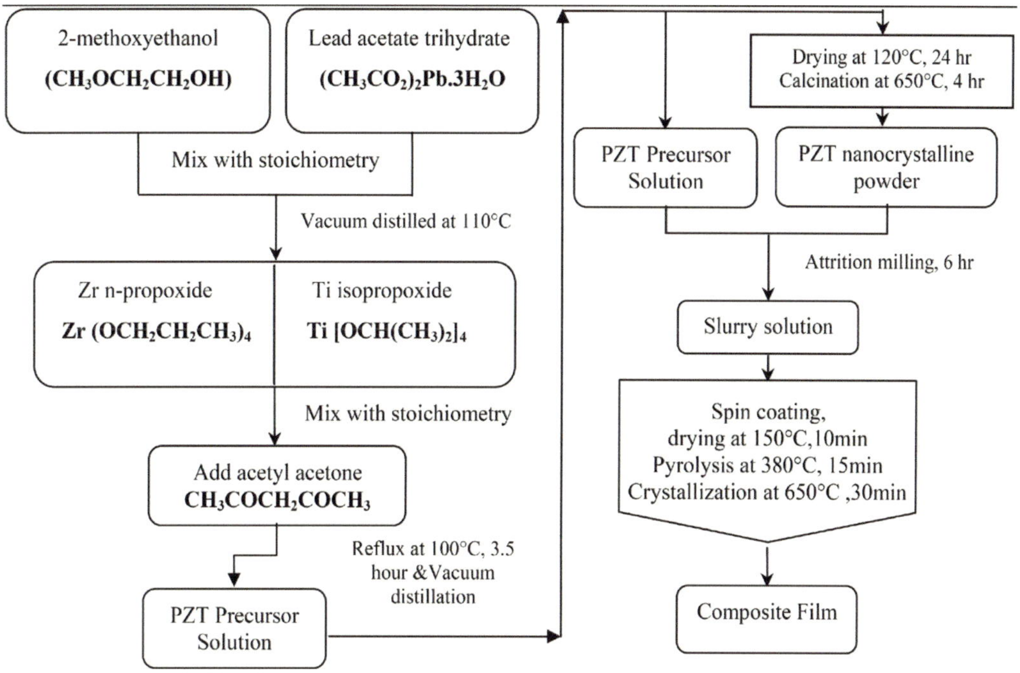

Figure 3. Flowchart of preparation PZT composite layer

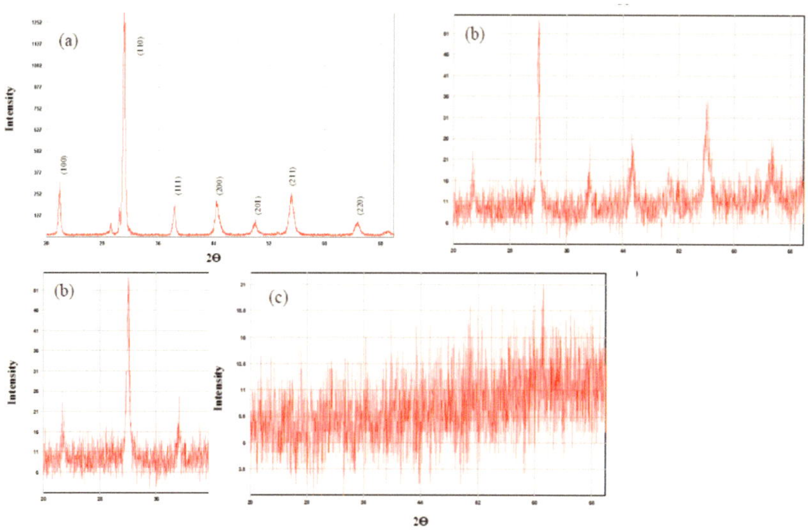

Figure 4. XRD patterns of: (a) PZT powders, (b) PZT Thin film with nanopowder, (c) PZT Thin film without nanopowders.

In Figure 5 the result of particle size analyzing is indicated. The powders from sol gel process are agglomerated and in range of 0.8μm.When these powders undergo attrition milling and some dispersant inserted to system, the size of particles reduces and after 6 hour a stable slurry will obtain. Increasing the time of milling has opposite effect and the particles become larger. This condition could attribute to increasing the temperature and re-agglomeration of particles. The average size of powders is 280nm.

Optical microscopic pictures of samples are shown in Figure 6. By increasing the amount of powder, the probability of crack existence in the film becomes smaller. This fact relates to formation of a strongly bonded network between sol gel film and ceramic particles. On the other hands less shrinkage in the films occurs due to the presence of powders and diminution of the percentage of sol gel in the film. However, the amount of powder greater than 50% leads to rough microstructure and porous films. The individual thickness of each layer can be increased through this technique and ferroelectric properties affected by this special microstructure.

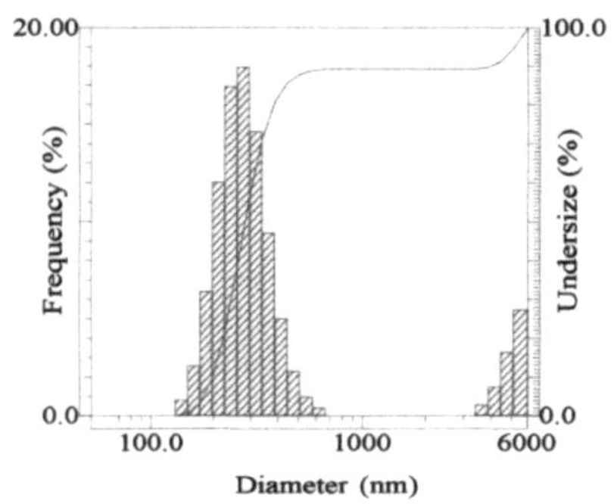

Figure 5. Particle size analyzing of a nanopowders

In Figure 7, the SEM images of films are shown. We can observe two kinds of grain in the films. One has irregular shape and larger grain size in order of 400-700nm; another one is granular and has a smaller grain size below 300nm.Particle loaded in slurry might be origin of the irregular shape and nucleation and growth in the sol solution forms the smaller grains group. Thus a composite island structure formed where the granular grains surrounding larger grains. This kind of microstructure will meet the requirement of both mechanical and electrical properties. From the morphologies, one can see that high density is achieved except few pinholes. However, microcracks with the length of microns distribute uniformly in the surface. This can be prevented by successful elimination of the aggregation among the nanopowders in the nanocomposite route.

4. Piezoelectric pressure sensors

The MEMS pressure sensors have been developed in the 1970's. Many works have been presented on developing these devices. For instance, Ravariu et al. [7] modeled a pressure

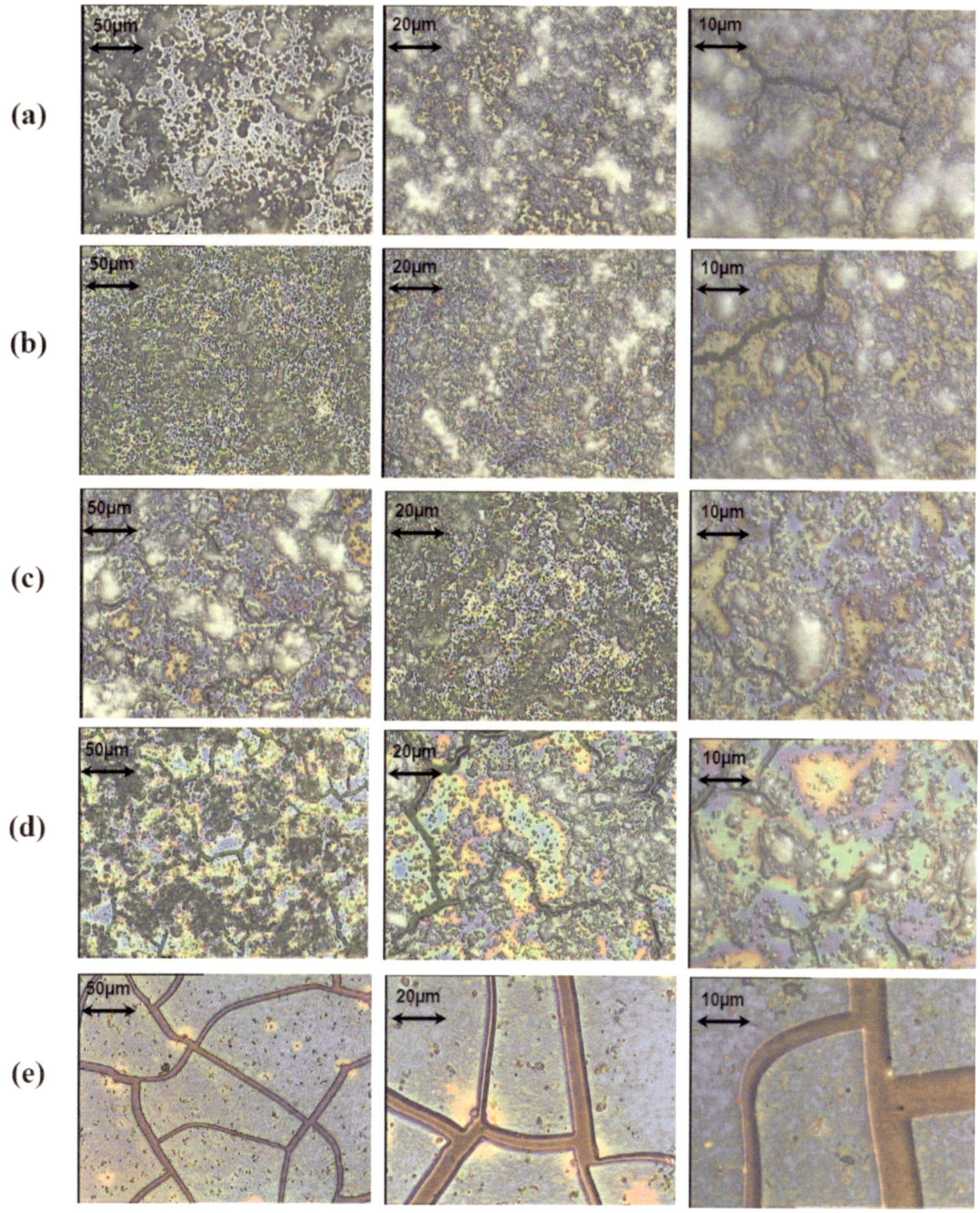

Figure 6. Optical microscopic pictures. Mass ratio of powder to sol is (a)$^{1}/_{1}$,(b) ¾,(c) ½,(d) ¼ (e) without powder

sensor for computing the blood pressure by using of ANSYS simulation in order to estimate the mechanical stress in their structure. Caliano et al. [60] described a new low-cost resonance piezoelectric sensor which uses, as the active element, a bimorph piezoelectric membrane, mass produced and widely employed in piezoelectric 'buzzers' and telephone receivers. The

Figure 7. SEM morphology of PZT composite films. Mass ratio of powder to solution: (a) ¼, (b) ½

membrane is driven by a very simple electronic circuit and provides as output a resonance frequency proportional, within a large range, to the applied pressure. Hindrichsen et al. [61] studied the effect of the thickness of the piezoelectric material on the sensor performance. Then compare the effect of thin and thick film piezoelectric materials on piezoelectric sensor performance. Mortet et al. [62] reported the frequency characterization of a commercially available piezoelectric bimorph microcantilevers and used as a pressure sensor. The cantilever operates as a driven and damped oscillator.

The PZT thin films are very attractive due to their larger piezoelectric properties compared to the most conventional piezoelectric materials. Hsu et al. [63] demonstrated an improved sol–gel process using rapid thermal annealing and a diluted sealant coating to obtain the PZT thickness of 2μm in three coatings with a crack-free area as large as 5mm×5mm. Then they characterized piezoelectric properties of the fabricated PZT films and demonstrated the use of the PZT thin film as a calibrated sensor. Luo et al. [64] presented the fabrication process for lead zirconate titanate (PZT) micro pressure sensor. The PZT piezoelectric thin film sensor is deposited on the surface of steel wafer at 300ºC by RF sputtering process and is micro fabricated directly. This PZT pressure sensor can be applied to on-line monitoring the pressure in the mold's core and cavity during injection process. Morten et al. [65] presented the study on the relationship between the composition, poling condition and piezoelectric properties of thick film layers. Microstructure, electrical and mechanical properties were analyzed. Pastes based on lead-titanate-zirconate (PZT) powders, with either PbO or a lead-alumina-silicate glass frit as binder, were prepared. Using this study the pressure sensor is described where a circular diaphragm of alumina supports two piezoelectric layers obtained by screen printing and firing a PZT/PbO-based ferroelectric paste. Liu et al. [5] designed two novel piezoelectric microcantilevers with two piezoelectric elements (bimorph or two segments of Lead Zirconate Titanate (PZT) films) and three electric electrodes. Morten et al. [66] described the design, implementation and performance of a resonant sensor for gas-pressure measurement realized with screen-printed and fired PZT-based layers on an alumina diaphragm.

http://dx.doi.org/10.5772/54755

5. Design and modeling procedure of a pressure sensor based on multilayer *ncpe*-PZT diaphragm

In this part and the following sections, the design and modeling procedure of a pressure sensor based on a square multilayer Si/SiO_2/Ti/Pt/ncpe-PZT/Au diaphragm by using of the ANSYS finite element (FE) software will be presented [72].

5.1. Design and modeling of the pressure sensor

The Cross section of the structure of the diaphragm is shown in Figure 8. Due to the symmetry of the diaphragm and in order to save some time and storage space in the calculation a quarter of the whole structure is used for simulation, as shown in Figure 9. The PZT film was deposited on Pt/Ti /SiO_2/Si wafer via sol-gel process. Au layer was evaporated on the surface of PZT film as a top electrode. The backside silicon was wet etched off till the SiO_2 layer.

Figure 8. Structure and Cross section of the multilayer PZT diaphragm.

5.2. Finite element analysis

The micro piezoelectric sensor was modeled in ANSYS® [67] FE software to obtain its electromechanical characteristics.

Three different element types were adopted to characterize different layers in the device. Solid 46 element type which has the capability of modeling the layered solid was used to model elastic layers of Pt, Ti and SiO_2. Using this element type helped modeling of three solid layers in one element layer. Piezoelectric layer was modeled using Solid5 element type. Substrate and Au layers were modeled using solid 45 element type. The small thickness-width ratio is the most challenging issue in the FE modeling of the thin films. Huge amount of elements are required to achieve suitable aspect ratio. In the present work, three layers of elements in modeling the membrane and substrate requires huge amount of CPU time. Some author used simplified model to decrease the CPU time [68], only modeled the diaphragm and substituted the surrounding membrane and substrate effect with clamped boundary condition. This simplified model is very rigid boundary condition. In this study a simplified model was used which assume that the substrate is a rigid body. The nodes in the interface between substrate and membrane were fixed. Modeling the thin film layers offers more relaxed boundary condition than one which is used in [68] despite its need to more CPU time.

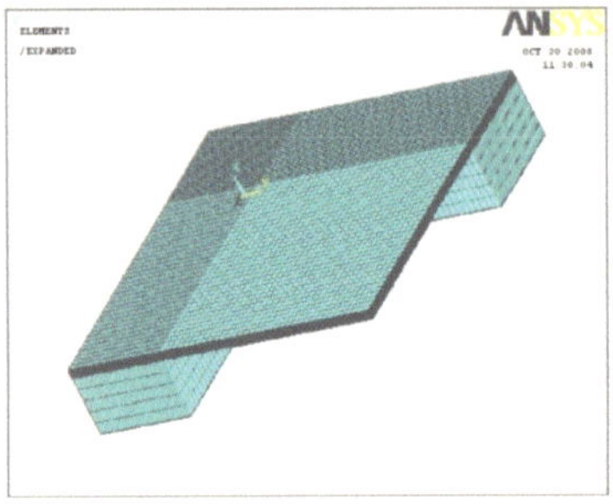

Figure 9. The model used in ANSYS which is a quarter of the whole structure with the regular meshes.

5.3. Simulation and optimization

Figure 10 shows the displacement, stress in different directions, total stress, and strain at 60mbar pressure. As shown in this figure the maximum displacement of the diaphragm is at the center, and the maximum stresses in different direction are along the edges of the diaphragm.

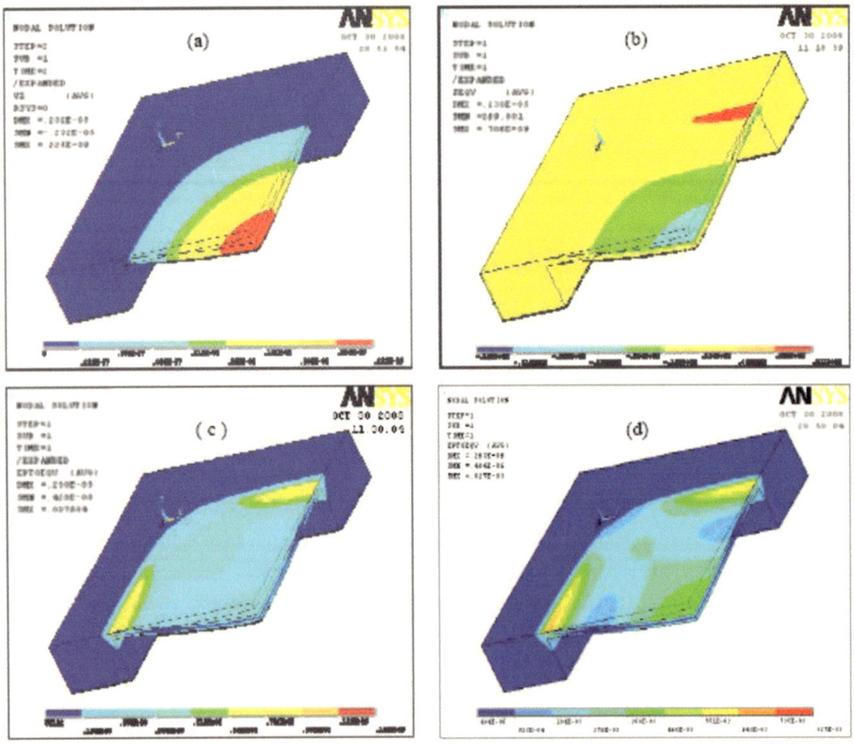

Figure 10. The simulation result from ANSYS (a) displacement; (b) stress in x direction; (c) stress total; (d) strain.

A finite element model of the device allowed us to investigate a dynamics characteristics and structural behaviors of the multilayer PZT diaphragm and optimize a PZT thin film diaphragm for use in the sensors and actuators applications.

Figure 11 shows the influence of the width of the square multilayer PZT diaphragm on the first nature frequency. The natural frequency decreases rapidly with increasing the diaphragm width especially in the larger PZT layer thicknesses. The x-axis of figure 10 was replaced with the reciprocal area of diaphragm, $1/a^2$, and is shown in the figure 12. As can be seen there is a linear relation between the 1st natural frequency and reciprocal area of diaphragm. Figure 13 shows the 1st natural frequency of the diaphragm regarding to the PZT layer thickness in the different diaphragm widths. There is also a linear relation between the 1st natural frequency and PZT layer thickness. Figures 12 and 13 can be used to predict the natural frequency in the larger models with more elements without further simulation. The dimension of the real MEMS sensors are larger than the sensor modeled in present work. There is three layers in the model, modeling a sensor which its dimension is 2 times larger than this model, require 12 times more elements than this model. Solving this huge amount of elements requires huge amount of CPU time to solve and hardware to save the model data.

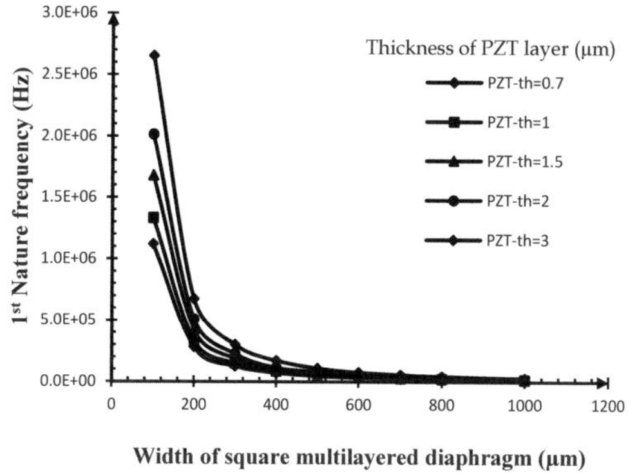

Figure 11. Change of the first nature frequency of structure via the variation of width of square laminated diaphragm

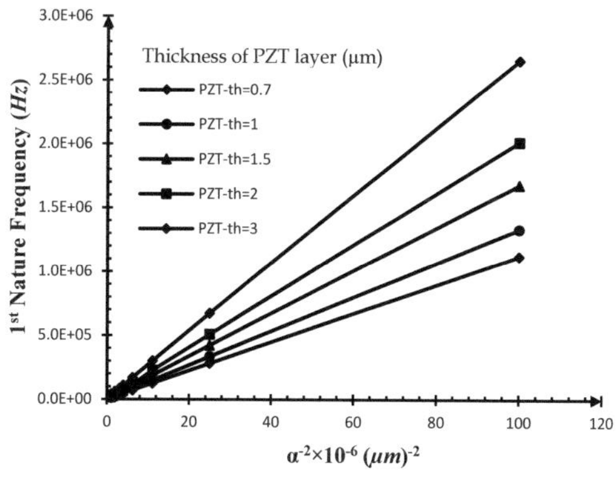

Figure 12. Change of the first nature frequency of structure via the inverse of square of width, $1/a^2$, of square laminated diaphragm

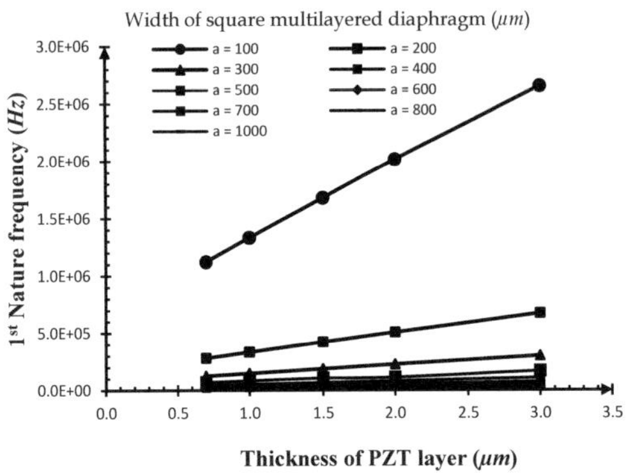

Figure 13. Effect of thickness of PZT-layer on the structure nature frequency

5.3.1. Harmonic response of the PZT thin film diaphragm

Harmonic response of the sensor was investigated under applying pressure load on the diaphragm. Since the supported silicon and surrounding multilayer film of diaphragm are not rigid solids and don't have clamped boundary, the exact model was adopted in investigation of the harmonic responses of the diaphragm. A damping ratio equal to 0.015 was assumed in the analysis. Figure 14 shows the deflection response at center point of the diaphragm. The voltage response PZT layer is shown in the figure 15. The dimensions of the exact model of a multilayer diaphragm which is modeled were a = 250μm, d = 750μm, h_1 = 200μm and h_5 = 1μm.

There is a little difference between the exact model and adopted simple model result and can be neglected. The exact model took CPU time 2.15 times more than simple model and can be used in further harmonic analysis in order to save the CPU time and reduce analysis cost.

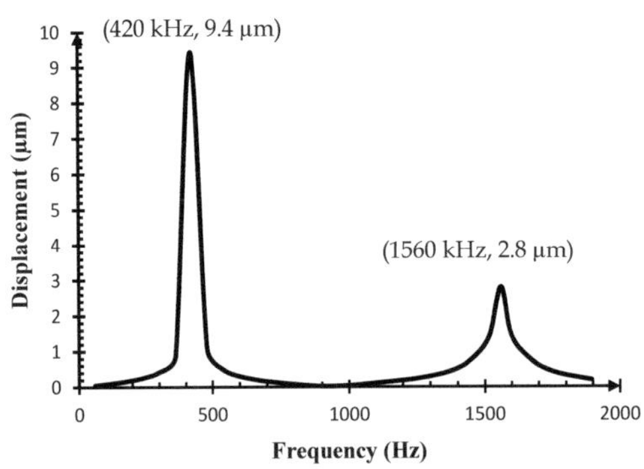

Figure 14. Displacement at centre point of the diaphragm

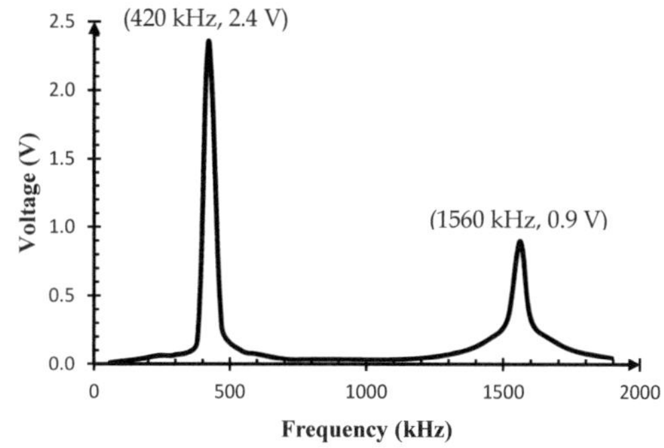

Figure 15. Voltage response in PZT-layer at centre point of the diaphragm

5.3.2. Effect of thickness ratio on dynamical behavior of the PZT thin film diaphragm

The thicknesses of SiO_2 and PZT are in the same order and because of it we considered variety of thickness ratios of them to study dynamical behavior of multilayer diaphragm for sensor and actuator structures are shown in figure 16. There exists an optimum thickness ratio of a PZT to elastic layers under which the deflection of the diaphragm will reach a maximum value. In the calculation, once the thickness of SiO_2 layer was varied, and the thickness of the PZT layer was remained at h_5 = 1μm, and then the thickness of PZT layer was varied, and the thickness of the SiO_2 layer was remained at h_2 = 400nm. The optimized values of thickness ratio for given widths versus the variations of PZT thickness for sensor and actuator structures are shown in Figure 17.

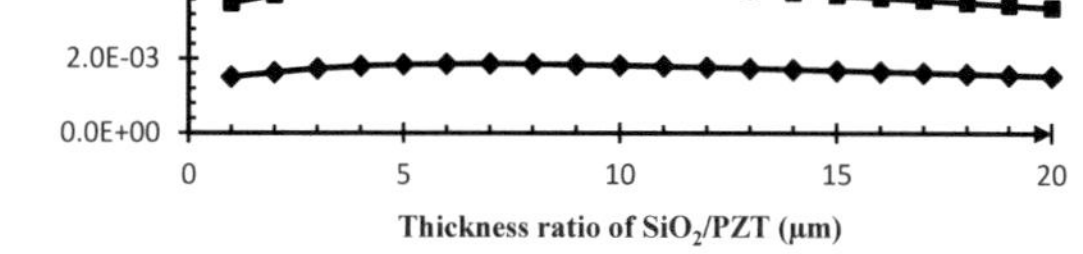

Figure 16. Dependence of the diaphragm deflection via the thickness ratio of SiO2/PZT layers of the diaphragm of a pressure sensor.

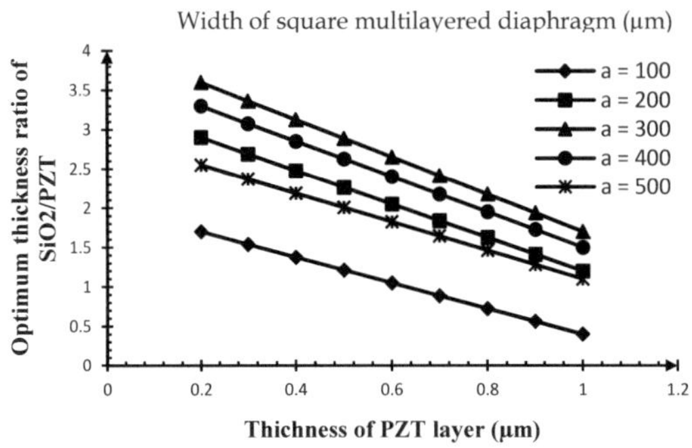

Figure 17. Change of optimum thickness ratio of SiO2/PZT plate versus variation of PZT thickness and diaphragm width of a pressure sensor

5.3.3. Sensor response

Figure 18 shows the voltage versus pressure diagram for different diaphragm widths. In this diagrams the PZT layer thickness is constant and equal to 1 µm and the SiO2 layer thickness is so selected to obtain the optimum value of R in that diaphragm width. These curves can be used to calibrate the sensor actuator response. There is a simple linear relation between voltage and pressure in each diaphragm width. This linear relation can be used to predict larger model behavior and calibrate them.

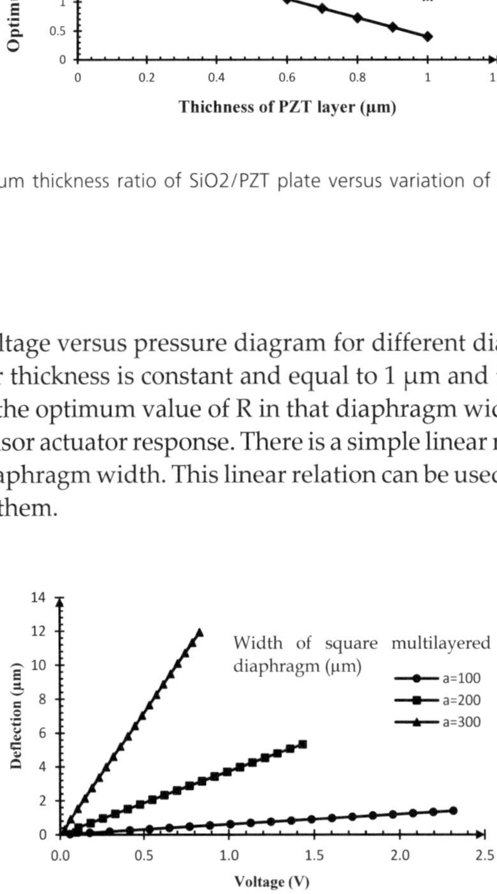

Figure 18. Diaphragm deflection versus voltage for different diaphragm widths.

6. Conclusions

In this chapter the nanocrystalline-powders-enhanced (*ncpe-*) PZT is introduced. This is the novel technique to enhance the piezoelectric properties of PZT sol-gel derived ceramics through the use of single crystal PZT microcubes as an inclusion in the PZT sol-gel. The preparation of the *ncpe*-PZT thin film have been presented. The deposited layer is characterized. The XRD analysis shows that perovskite structure would be formed due to the presence of a significant amount of ceramic nanopowders and the (100) preferred direction was induced. This texture has a good effect on piezoelectric properties of perovskite structure. The film forms a strongly bonded network and less shrinkage occurs, so the films do not crack during process. The aspect ratio through this process would be increased. SEM micrographs indicated that *ncpe*-PZT films were uniform, crack free and have a composite microstructure. The dynamics characteristics and structural behaviors of the multilayer *ncpe*-PZT diaphragm were investigated in the pressure sensor structure. The effective parameters of the multilayer PZT diaphragm for improving the performance of a pressure sensor in different ranges of pressure are optimized. The influence of dimensions of multilayer diaphragm on nature frequency was studied. The frequency values were found to rapidly decrease with the increase in the diaphragm width, especially in the small width range, and to increase linearly with the increase in the thickness of the *ncpe*-PZT-layer. The deflection and the first nature frequency of diaphragm as a function of the thickness ratio of *ncpe*-PZT layer to SiO_2 layer were presented for the optimum design of actuator or sensors in MEMS applications.

By information given by this chapter, the readers become more familiar with piezoelectric ceramic materials, piezoelectric properties and characterization followed by the techniques of the thin-film processing. The readers are also introduced by preparation and thin-film deposition steps of the nanocrystalline-powders-enhanced (*ncpe-*) PZT layers with the steps of the design, modeling and optimizing of the pressure sensor based on the *ncpe*-PZT thin-films multilayer diaphragm. The results give the good tool for the people who wants to use the PZT for the diaphragm based MEMS sensors.

Acknowledgements

The authors would like to thank Prof. M. H. Sheikhi, Prof. S. Mohajerzadeh, Dr. A. Barzegar, Dr. E. Masumi and Mr. S. Torkian (MSc) and the staff of the Nanotechnology Research Institute (NRI) of Shiraz University, Thin Film Lab. of Tehran University and Thin Film Lab. of the Dept. of Physics of Shiraz University. And thanks to all people who participated in this work.

Author details

Vahid Mohammadi[1], Saeideh Mohammadi[2] and Fereshteh Barghi[3]

1 Delft Institute of Microsystems and Nanoelectronics (dimes), Delft University of Technology, The Netherlands

2 Isfahan University of Technology, Isfahan, Iran

3 Shiraz University, Shiraz, Iran

References

[1] Proceeding of the International Conference of Solid-State Sensors and ActuatorsJune, (1997). Proc. IEEE Workshop of Micro Electro Mechanical Systems, Heidelberg, Germany, January 1998.

[2] Gabriel, K. J. Engineering Microscopic Machines. Scientific American (1995). , 273(3), 150-153.

[3] Microelectromechanical Systems-A DoD Dual Use Technology Industrial AssessmentU.S. Dept. Defense Final Report. (1995).

[4] Sze, S. M. Semiconductor Sensors. New York: Wiley; (1994).

[5] Liu, M, Cui, T, Dong, W, Cuil, Y, Wang, J, Du, L, & Wang, L. Piezoelectric microcantilevers with two PZT thin film elements for microsensors and microactuators. Proceedings of the 1st IEEE International Conference on Nano/Micro Engineered and Molecular Systems, Zhuhai, China, (2006).

[6] Zinck, C, Pinceau, D, Defay, E, Delevoye, E, & Barbier, D. Development and characterization of membranes actuated by a PZT thin film for MEMS applications. Sensors and Actuators A (2004). , 115, 483-489.

[7] Ravariu, F, Ravariu, C, & Nedelcu, O. The modeling of a sensor for the human blood pressure. International IEEE Semiconductor Conference, (2002). CAS 2002 Proceeding , 1, 67-70.

[8] Polla, D. L, & Francis, L. F. processing and characterization of piezoelectric materials and integration into microelectromechanical systems. Annual review Material Science (1998). , 28, 563-597.

[9] Jaffe, B, Cook, W, & Jaffe, H. Piezoelectric Ceramics. New York: Academic; (1971).

[10] Nye, J. F. The Physical Properties of Crystals. London: Oxford Univ. Press; (1957).

[11] Cady, W. G. Piezoelectricity. New York: Dover; (1962).

[12] Abrahams, S. C, & Nassau, N. In Concise Encyclopedia of Advanced Ceramic Materials, ed. RJ Brook. Cambridge, MA: MIT Press; (1991). , 351-354.

[13] Lines, M. E, & Glass, A. M. Principles and Applications of Ferroelectrics and Related Materials. Oxford: Clarendon; (1977).

[14] Forster, N. F. Handbook of Thin Film Technology, ed. Maissel LI, Glang R, New York: McGraw-Hill; (1970). Chapter. 15.

[15] Polla, D. L, Muller, R. S, & White, R. M. Integrated multisensor chip. IEEE Electron Device Letter (1986). , 7(4), 254-256.

[16] Heywang, W, & Thomann, H. Tailoring of Piezoelectric Ceramics. Annual Review Material Science (1984). , 14, 27-47.

[17] Hardtl, K. H. Physics of ferroelectric ceramics used in electronic devices. Ferroelectrics (1976)., 12, 9-19.

[18] Cross, L. E, & Hardtl, K. H. Ferroelectrics. In Encyclopedia of Chemical Technology. NewYork: Wiley&Sons; (1980)., 10

[19] Choi, S. W, Shrout, T. R, Jang, S. J, & Bhalla, A. S. Morphotropic phase boundary in Pb(Mg1/3Nb2/3)O3-PbTiO3 system. Material Letters (1989)., 8, 253-55.

[20] Nomura, S, Yonezawa, M, Doi, K, Nanamatsu, S, Tsubouchi, N, & Takahashi, M. NEC Res. Dev (1973)., 29, 15-21.

[21] Yamashita, Y. Improved Ferroelectric Properties of Niobium-Doped $Pb[(Sc_{1/2}Nb_{1/2})Ti]O_3$ Ceramic Material. Jpn. J. Appl. Phys. (1993)., 32, 5036-5040.

[22] Haertling, G. H. Ceramic Materials for Electronics: Processing, Properties, and Applications, ed. Buchanon RC. New York: Dekker; (1986)., 139-226.

[23] Scott, J. F. Paz Araujo CA. Ferroelectric memories. Science (1989)., 246, 1400-1405.

[24] Tarui, Y. Technical Digest-Int. Electron Devices Meet. Piscataway, NJ: IEEE, (1994)., 7-16.

[25] Dimos, D, Lockwood, S. J, Garino, T. J, Al-shareef, H. N, & Schwartz, R. W. Ferroelectric Thin Films V, MRS Symp. Proc. 433. ed. SB Desu, R Ramesh, BA Tuttle, RE Jones, IK Yoo. Pittsburg, PA: Mater. Res. Soc. (1996)., 305-16.

[26] Myers, E. R, & Kingon, A. I. Ferroelectric Thin Films, MRS symposium proceeding 200, Ferroelectric Thin Films. (1990)., 141-152.

[27] Kingon, A. I, Myers, E. R, & Tuttle, B. Ferroelectric Thin Films II, MRS Symp. Proc. 243. Pittsburg: Mater Res. Soc., (1992).

[28] Desu, S. B, Ramesh, R, Tuttle, B. A, Jones, R. E, & Yoo, I. K. Ferroelectric Thin Films V, MRS Symp. Proc. 433. Pittsburg: Mater. Res. Soc., (1996).

[29] Auciello, O, Kingon, A. I, & Krupanidhi, S. B. Sputter Synthesis of Ferroelectric Films and. Heterostructures. MRS Bulletin (1996)., 21, 25-30.

[30] De Keijser, M. Dormans GJM. Chemical vapor deposition of electroceramic thin films. MRS Bulletin (1996)., 21, 37-43.

[31] Lisa, C. Klein, Sol-Gel Technology for Thin Films, Fibers, Preforms, Electronics and Specialty Shapes, Publisher: William Andrew; 1ed (1989).

[32] Cooney, T. G, & Francis, L. F. Processing of sol-gel derived PZT coatings on non-planar substrates. Journal of Micromechanics and Microengineering (1996)., 6, 291-300.

[33] Francis, L. F. Sol-Gel Methods for Oxide Coatings. Journal of Materials Processing Technology (1997)., 12, 963-1015.

[34] Brinker, C. J, & Scherer, G. W. Sol-Gel Science: The Physics and Chemistry of Sol-Gel Processing. New York: Academic; (1990).

[35] Tuttle, B. A, & Schwartz, R. W. Solution deposition of ferro- electric thin films. MRS Bulletin 21, 49-54, (1996).

[36] Yi, G, & Sayer, M. Sol-gel Processing of Complex Oxide Films. Ceramic Society Bulletin (1991). , 70, 1173-1179.

[37] Aegerter, M. A. Ferroelectric thin coatings. Journal of Non-Crystalline Solids (1992). , 151, 195-202.

[38] Wilkinson, A. P, Speck, J. S, Cheetam, A. K, Natarajan, S, & Thomas, J. M. In situ x-ray diffraction study of crystallization kinetics in $PbZr_xTi_xO_3$, (PZT, x = 0.0, 0.55, 1.0). Chemistry of Materials 6(6), 750-754, (1994). , 1.

[39] Tuttle, B. A, Schwartz, R. W, Doughty, D. H, & Voight, J. A. Ferroelectric Thin Films, MRS Symp. Proc. 200. Pittsburg: Mater. Res. Soc. (1990). , 159-65.

[40] Tuttle, B. A, Headley, T. J, Bunker, B. C, Schwartz, R. W, Zender, T. J, Hernandez, C. L, Goodnow, D. C, Tissot, R. J, Michael, J, & Carim, A. H. Microstructural evolution of $Pb(Zr, Ti)O_3$ thin films prepared by hybrid metallo-organic decomposition. Journal of Materials Research 7(7), 1876- 1882, (1992).

[41] Hsueh, C. C, & Meartney, M. L. Microstructural Evolution of Sol-Gel Derived PZT Thin Films. MRS Proceedings 243, 451, (1991).

[42] Kwok, C. K, & Desu, S. B. Formation kinetics of PZT thin films. Journal of Materials Research 9(7), 1728-1733, (1994).

[43] Lefevre, M. J, Speck, J. S, Schwartz, R. W, Dimos, D, & Lockwood, S. J. Microstructural development in sol-gel derived lead zirconate titanate thin films: The role of precursor stoichiometry and processing environment. Journal of Materials Research 11(8), 2076-2084, (1996).

[44] Bernstein, S. D, Kisler, Y, Wahl, J. M, Bernacki, S. E, & Collins, S. R. Ferroelectric Thin Films II, MRS Symp. Proc. 243. Pittsburg: Mater Res. Soc., (1992). , 373-378.

[45] Wright, J. S, & Francis, L. F. Phase development in Si modified sol-gel-derived lead titanate Journal of Materials Research 8(7), 1712-1720, (1993).

[46] Francis, L. F, & Payne, D. A. Thin-Layer Dielectrics in the $Pb[(Mg_{1/3}Nb_{2/3})_{1-x}Ti_x]O_3$ System. Journal of the American Ceramic Society (1991). , 74, 3000-3010.

[47] Lakeman CDE, Xu Z, Payne DA. Rapid thermal processing of sol-gel derived PZT 53/47 thin layers. Proc. 9th IEEE Int. Symp. Applications of Ferroelectrics, ISAF' (1994). ., 94, 404-407.

[48] Griswold, E. M, Weaver, L, Sayer, M, & Calder, I. D. Phase transformations in rapid thermal processed lead zirconate titanate. Journal of Materials Research 10(12), 3149-3159, (1995).

[49] Chen, S. Y, & Chen, I-W. Temperature-Time Texture Transition of $Pb(Zr_{1-x}Ti_x)O_3$ Thin Films: I, Role of Pb-rich Intermediate Phases (and Temperature-Time Texture Transi-

tion of $Pb(Zr_{1-x}Ti_x)O_3$ Thin Films: II, Heat Treatment and Compositional Effects (pages 2337-2344), Journal of the American Ceramic Society 77, 2332-2336 and 2337-2344, (1994)., 2332-2336.

[50] Hren, P. D, Rou, S. H, Al-shareef, H. N, Ameen, M. S, Auciello, O, & Kingon, A. I. Bottom electrodes for integrated $Pb(Zr, Ti)O_3$ films. Integrated Ferroelectrics (1992)., 2, 311-325.

[51] Sreenivas, K, Reaney, I, Maeder, T, Setter, N, Jagadish, C, & Elliman, R. G. Investigation of Pt/Ti bilayer metallization on silicon for ferroelectric thin film integration. Journal of Applied Physics (1994)., 75, 232-239.

[52] Bruchhaus, R, Pitzer, D, Eibl, O, Scheithuer, U, & Joesler, . . Ferroelectric Thin Films II, MRS Symp. Proc. 243. Pittsburg: Mater Res. Soc. 123-128, 1992.

[53] Al-shareef, H. N, Kingon, A. I, Chen, X, Ballur, K. R, & Auciello, O. Contribution of electrodes and microstructures to the electrical properties of $Pb(Zr_{0.53}Ti_{0.47})O_3$ thin film capacitors. Journal of Materials Research 9(11), 2968-2975, (1994).

[54] Tuttle, B. A, Garino, T. J, Voight, R. A, Headley, T. J, Dimos, D, & Eatough, M. O. Science and Technology of Electroceramic Thin Films, ed. O Auciello, R Waser, The Netherlands: Kluwer; (1995)., 117.

[55] Lefki, K. Dormans GJM. Measurement of piezoelectric coefficients of ferroelectric thin films. Journal of Applied Physics (1994)., 76, 1764-1767.

[56] Zhang, Q. M, Pan, W. Y, & Cross, L. E. Laser interferometer for the study of piezoelectric and electrostrictive strains. Journal of Applied Physics (1988)., 63, 2492-2496.

[57] Li, J. F, Moses, P, & Viehland, D. simple, high-resolution interferometer for the measurement of frequency-dependent complex piezoelectric responses in ferroelectric ceramics. Review of Scientific Instruments (1995)., 66, 215-221.

[58] Tsaur, J, Wang, Z. J, Zhang, L, Ichiki, M, Wan, J. W, & Maeda, R. Preparation and Application of Lead Zirconate Titanate (PZT) Films Deposited by Hybrid Process: Sol-Gel Method and Laser Ablation. Japanese Journal of Applied Physics (2002). , 41, 6664-6668.

[59] Ong, R. J, Berfield, T. A, Sottos, N. R, & Payne, D. A. Sol-gel derived $Pb(Zr,Ti)0_3$ thin films: Residual stress and electrical properties. Journal of the European Ceramic Society (2005)., 25, 2247-2251.

[60] Caliano, G, Lamberti, N, Lula, A, & Pappalardo, M. A piezoelectric bimorph static pressure sensor. Sensors and Actuators A 46(1-3): 176-178, (1995).

[61] Hindrichsen, C. G, Lou-møller, R, Hansen, K, & Thomsen, E. V. Advantages of PZT thick film for MEMS sensors. Sensors and Actuators A (2010)., 163, 9-14.

[62] Mortet, V, Petersen, R, Haenen, K, & Olieslaeger, D. M. Wide range pressure sensor based on a piezoelectric bimorph microcantilevers. Applied Physics Letters 88, 133511, (2000).

[63] Hsu, Y. C, Wu, C. C, Lee, C. C, Cao, G. Z, & Shen, I. Y. Demonstration and characterization of PZT thin-film sensors and actuators for meso- and micro-structures. Sensors and Actuators A (2004). , 116, 369-377.

[64] Luo, R. C, & Chen, C. M. PZT Thin Film Pressure Sensor for On-line Monitoring Injection Molding. 26th Annual IEEE Conference of the Industrial Electronics Society (IECON) (2000). , 4, 2394-2399.

[65] Morten, B, Cicco, G. D, Gandolfi, A, & Tonelli, C. PZT-based Thick Films and the Development of a Piezoelectric Pressure Sensor. Microelectronics International 9(2), 25-28, (1992).

[66] Morten, B, Cicco, G. D, & Prudenziati, M. Resonant pressure sensor based on piezoelectric properties of ferroelectric thick films. Sensors and Actuators A (1992). , 31, 153-158.

[67] ANSYS guide. http://www.ansys.com.

[68] Yao, L. Q, & Lu, L. Simplified Model and Numerical Analysis of Multi-layered Piezoelectric Diaphragm. Advanced Materials for Micro- and Nano- Systems (AMMNS), (2003).

[69] Mohammadi, V, Sheikhi, M. H, Torkian, S, Barzegar, A, Masumi, E, & Mohammadi, S. Design, modeling and optimization of a piezoelectric pressure sensor based on thin-film PZT diaphragmcontain of nanocrystalline powders. 6th International Symposium on Mechatronics and its Applications (ISMA'09), 1-7, (2009).

[70] Lin, Y, Andrews, C, & Sodano, H. A. Enhanced piezoelectric properties of lead zirconate titanate sol-gel derived ceramics using single crystal $PbZr_{0.52}Ti_{0.48}O_3$ cubes. Journal of Applied Physics 108(6), 064108, (2010).

[71] Luo, C, Cao, G. Z, & Shen, I. Y. Enhancing displacement of lead-zirconate-titanate (PZT) thin-film membrane microactuators via a dual electrode design. Sensors and Actuators, A 173(1), 190-196, (2012).

[72] Mohammadi, V, Masoumi, E, Sheikhi, M. H, & Barzegar, A. Design, Modeling and Optimization of a Multilayer Thin-Film PZT Membrane Used in Pressure Sensor. Third international conference on modeling, simulation, and applied optimization (ICM-SAO'09) january 20-22, (2009). sharjah, UAE.

Chapter 7

Design and Application of Piezoelectric Stacks in Level Sensors

Andrzej Buchacz and Andrzej Wróbel

Additional information is available at the end of the chapter

http://dx.doi.org/10.5772/54580

1. Introduction

In recent years there is growing interest of materials, called smart materials. They have one or more properties that can be significantly changed. Smartness describes abilities of shape, size and state of aggregation changes. The main groups of smart materials are:

- piezoelectric plates,
- magneto-rheostatic materials,
- electro-rheostatic materials,
- shape memory alloys.

Those materials are widely used in technology and their numbers of applications still growing. Piezoelectric effect was discovered by French physicists Peter and Paul Curie in the 1880s. They described generation of electric charge on the surface with various shape during its deformation in different directions.

In their research, first of all, they focused on tourmaline crystal, salt and quartz. In 1881s Gabriel Lippman suggested the existence of the reverse piezoelectric phenomenon, which was confirmed experimentally by the Curie brothers. As a solution of research such two, unique properties of piezoelectric materials were assigned:

- showing of simple piezoelectric effect, which rely on generating of voltage after deformation of material,
- reverse piezoelectric effect, which rely on changing of sizes (by around 4%) after applying a voltage to piezoelectric facing.

Designing of technological systems, which contains piezoelectric elements should not be framed only to mechanical system analysis, but should be taken under consideration also electrical part. The entity should be considered as complex system, which contains independent subsystem.

Problem with mechanical-physical systems synthesis, first of all electrical and mechanical ones, is well known and frequently published (Arczewski, 1988; Bellert 1981; Białas, 2012; Buchacz & Płaczek, 2012). In articles concerned theory and designing of filters not much space was devoted to mechanical systems with parameters distributed in continuous way. Determination tests of mechatronic systems characteristics, applications of graphs and structural numbers were carried out at Silesian Center repeatedly (Buchacz, 2004; Sękala & Świder, 2005; Wróbel, 2012). Those studies gave assumption to analysis of piezoelectric work. In many publications and papers, mechanical systems investigations on example of vibration beams and rods, were introduced. Moreover, rules of modellinig by non-classical method and attempts of analysis by using hypergraph skeletons (Buchacz & Świder, 2000), graphs with signal flow and matrix methods (Bishop et at.,1972) were shown in these works.

Nowadays, numerous piezoelectric advantages caused its multi-application in mechanics and in many replaced field of science (Shin et al.,2005; Ha, 2002). Many times beams configurations, with respect to different boundary conditions and during piezoelectric application in damping of vibrations, were analyzed. In the paper (Sherrit, 1999) capability of piezoelectric systems modelling using equivalent Manson models were presented. Analysis of longitudinal vibrations were made taking into consideration dielectric and piezoelement layer. Mason in (Mason, 1948) introduced one-dimensional, equivalent system parameters widely used in modelling systems both free, and loaded. The main disadvantage of such approach is the equivalent of the mechanical system by discreet model. In article (Shin et at.,2005) author presents 4-port equivalent system of piezoelectric plate, used to identification of system response on mechanical force. A matrix, size 5x5, input-output dependences, with different conditions of support, was also determined. Another type of piezoelectric transducer, which was based on Masons alternative systems of higher number of piezoelectric layer, were presented in many articles. Simulation was carried out in frequency domain, furthermore result was compared with values obtained by experimental method. Bellert in his volumes of chosen works (Bellert 1981) many times wrote about modelling of replacing systems, examined as 4-ports. In work (Bolkowski, 1986) author provide chain method of connection electric 4-ports. However, both: (Bolkowski, 1986) and (Bellert 1981) concerned primarily electric systems. In research work number N502 071 31/3719 attempts of active, mechanical systems, with damping in scope of graphs and structural numbers methods, were analyzed. In such, rich publications from field of vibration analysis, solution of piezoelectric plate itself with respect to dynamic characteristic was not undertaken, with the exceptions (Kacprzyk, 1995).

Previous presented solutions were conducted mainly in field of time and concerned single plate. Present paper is continuation of mentioned publications with stack of piezoelectric plates. This work is an author's idea of calculations of complex systems with many elements. The base of calculation is matrix method and application of aggregation of graphs to determination characteristic parameters of bimorphic systems, as well as to drawing its characteristics.

http://dx.doi.org/10.5772/54580

2. Vibrating level sensor as a practical example of the application of piezoelectric stacks

An important characteristic of the designed and analyzed piezoelectric systems is the possibility of their practical application. This chapter presents options for further research related to the piezoelectric phenomenon of complex systems. Both a single piezoelectric plate, as well as complex systems, are often used in pressure, level, force and displacement sensors. As part of future research is proposed execution of laboratory stand for tests of piezoelectric plates used in vibration sensors. These sensors were used for level detection of loose materials in open or pressurized tanks. Output signal is a binary signal, transmitted to the automation systems via a relay. In Fig. 1 and Fig. 2 the level of vibration sensors manufactured by the "Nivomer" company from Gliwice were shown.

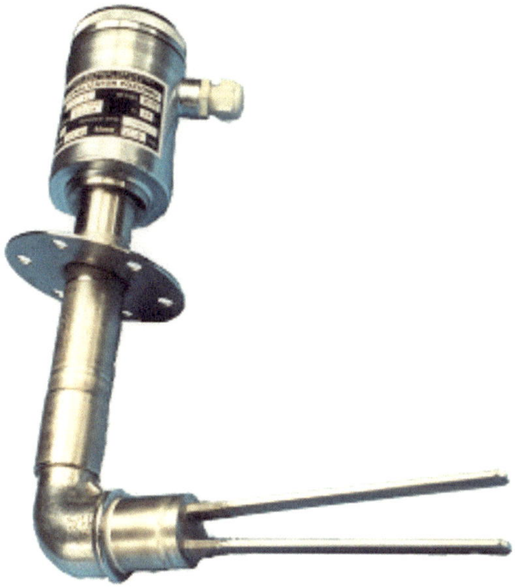

Figure 1. Angular vibrational sensor

The application of stack tiles for intensification of the output signal. The sensors consist of two pairs of receiving plates and two or three supplying plates, connected in a bimorphic system (Fig. 3, Fig. 4). Variable voltage, which feeds the supply plates results in a change in their thickness proportionally to the value of applied voltage.

Changes in the plate thickness causing mechanical vibrations of the element, so called "fork". When the "forks" are not covered by material, full deformation of supplying plates are transferred to the receiving ones. As a result of elongation of receiving plates, on its facing, there is a difference of potentials, proportional to the force. The value of this voltage is

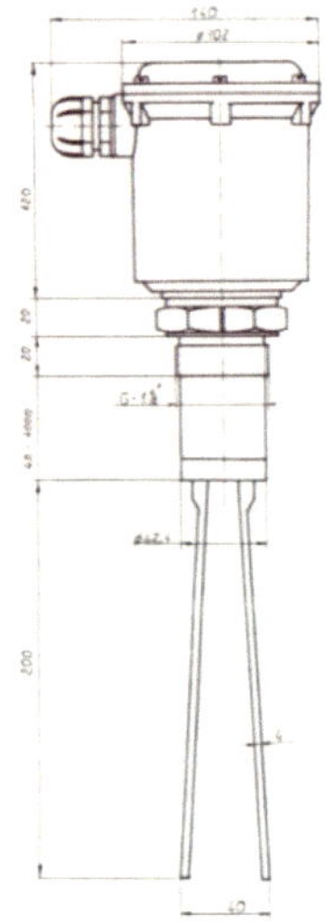

Figure 2. Approximate dimensions [documentation of the "Nivomer"company]

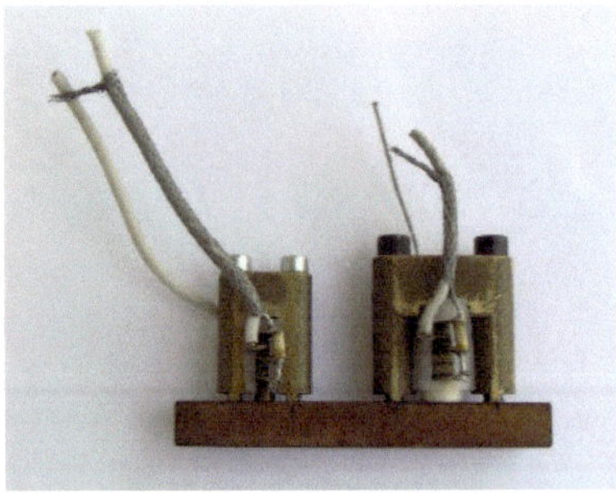

Figure 3. Stack of plates in the level sensor

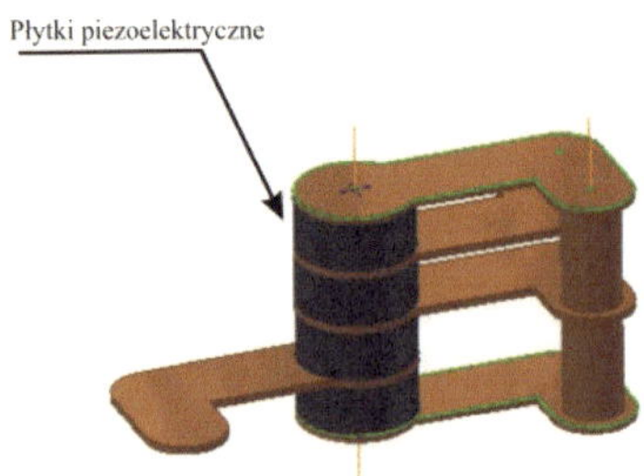

Figure 4. Computer model of the tiles stack

transformed by an electronic system (Fig. 5). In case of covered "forks", the receiving plates are no longer crushed and stretched. At the same time the potential is not generated on the facing of the plates.

http://dx.doi.org/10.5772/54580

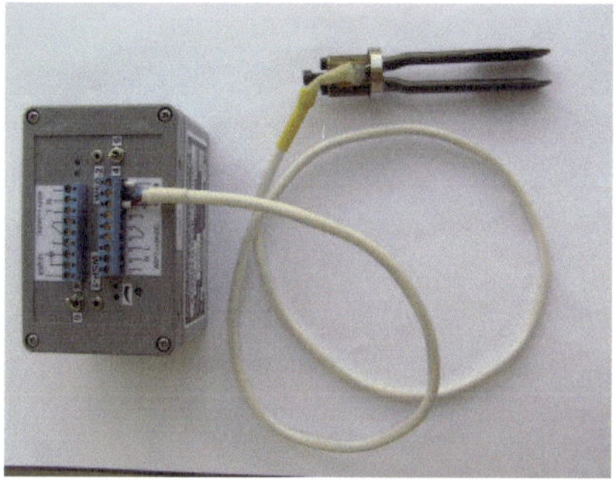

Figure 5. View of the sensor with control

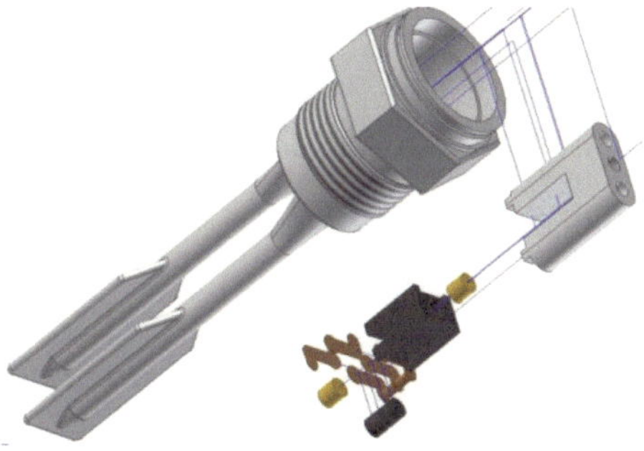

Figure 6. Construction of the level sensor by "Nivomer" company

Described sensor of "Nivomer" company is made of the body, ended with membrane which intensify a signal, to which are welded two identical vibrating rods (Fig. 6). In the presented sensor piezoelectric plates in the form of discs were used but there is a possibility of replacing the plates with shapes analyzed in previous chapters. Effect of plates stacks analysis in sensors designing The proposed analysis of the piezoelectric phenomenon of bimorphic systems allows at the design stage to determine the optimal parameters of piezoelectric plates. Well-chosen plate size and their number is crucial in the performance of other mechanical parts.

After the preliminary analysis of the construction of sensors with the company "Nivomer" from Gliwice found that by introducing a variable number of plates in the system is possible to choose the frequency of the generator and the maximum deflection of the fork carrying vibrations. The proposed methods and algorithms of work concern complex systems, can be used to design a stack of piezoelectric plates in the presented sensor level. In future work, it is proposed to conduct vibration test level sensors and a comparison of the algebraic method with the experimental method.

3. Object model under examination

Under consideration is vibrating piezoelectric plate with parameters distributed in a continuous way. The model has a section A, a thickness d and is made of a uniform material with a density ρ. Example of such system was shown in Fig. 7. Mechanical displacements are caused by the forces and voltage, while the current is generated by the difference of potentials on the plates of piezoelectric.

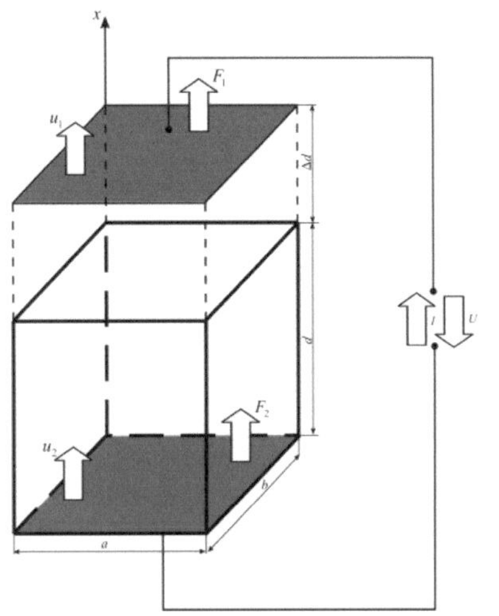

Figure 7. Continuous and limited piezoelectric model

In the analyzed example calculations are based on constitutive equations that include the assumed boundary conditions. In the study assumed that the test object is vibrating piezoelectric plate treated as a one-dimensional system. Piezoelectric plate constitutive equations are as follows (Soluch, 1980):

$$\begin{cases} \sigma = E\dfrac{\partial u}{\partial x} - \varepsilon E_p, \\ D = \varepsilon^S E_p + \varepsilon \dfrac{\partial u}{\partial x}, \end{cases} \tag{1}$$

where:

E - the modulus of longitudinal elasticity,

E_p - the intensity value of electric field,

ε - deformation,

ε^S - the electric permeability,

D - the electric induction.

The equation of motion of a given element is as follows:

$$\frac{\partial \sigma}{\partial x} = \rho \ddot{u}, \tag{2}$$

where:

σ - tension,

ρ - density of piezoelectric plate.

It was assumed that Poisson equation is:

$$\frac{\partial D}{\partial x} = 0. \tag{3}$$

Rearranging equation (1) due to E_p :

$$E_p = \frac{D}{\varepsilon^S} - \frac{\varepsilon}{\varepsilon^S} \frac{\partial u}{\partial x}. \tag{4}$$

Substituting expression (4) into equation (1), strain is given:

$$\sigma = c \frac{\partial u}{\partial x} - \frac{\varepsilon}{\varepsilon^S} D, \tag{5}$$

where:

$c = E + \varepsilon^2 / \varepsilon^S$ is the stiffened elastic constant.

From equation (3) result that $D = const$.Therefore, the equation of motion (2), that takes into account (5), is a one-dimensional wave equation:

$$c \frac{\partial^2 u}{\partial x^2} = \rho \ddot{u}, \tag{6}$$

or assuming that the volume wave equation in piezoelement is equal to:

$$V = \sqrt{\frac{c}{\rho}}, \tag{7}$$

equation of motion (6) was presented in the form:

$$\frac{\partial^2 u}{\partial x^2} - \frac{1}{V^2}\ddot{u}. \tag{8}$$

Assuming the expansion of the plate, mainly in the perpendicular plane to the axis, the following boundary conditions were defined:

$$\begin{cases} u = u_1, & when \quad x = x_1, \\ u = u_2, & when \quad x = x_2, \\ \sigma = -\sigma_1, & when \quad x = x_1, \\ \sigma = -\sigma_2, & when \quad x = x_2. \end{cases} \tag{9}$$

Replaced mechanical stress by force, using the formula $F = A\sigma$, where A is the surface of piezoelectric plate and σ is the stress of piezoelectric plate. Determined forces in this case are:

$$\begin{cases} F = F_1, & when \quad x = x_1, \\ F = F_2, & when \quad x = x_2. \end{cases} \tag{10}$$

The solution of equation (8) is harmonical displacement:

$$u = a_1 e^{j(\omega t - kx)} + a_1 e^{j(\omega t + kx)}, \tag{11}$$

or assuming, that

$$A_1 = a_1 e^{j\omega t},\ A_2 = a_2 e^{j\omega t},\ A_2 = a_2 e^{j\omega t}. \tag{12}$$

written as:

$$u = A_1 e^{-jkx} + A_2 e^{jkx}. \tag{13}$$

Taking into account the boundary conditions included in the model (9) was obtained:

$$\begin{cases} u_1 = A_1 e^{-jkx_1} + A_2 e^{jx_1}, \\ u_2 = A_1 e^{-jkx_2} + A_2 e^{jx_2}. \end{cases} \tag{14}$$

In order to determine the coefficients A_1 and A_2, presented in equation (12) the determinants method was used. The determinant of the set of equations (14) takes the form:

$$W = \begin{vmatrix} e^{-jkx_1} & e^{jkx_1} \\ e^{-jkx_2} & e^{jkx_2} \end{vmatrix}, \tag{15}$$

$$|W| = e^{-jkx_1} e^{jkx_2} - e^{jkx_1} e^{-jkx_2}, \tag{16}$$

$$|W| = e^{-jkx_1 + jkx_2} - e^{-jkx_2 + jkx_1}. \tag{17}$$

Using the dependences between the trigonometric and exponential functions:

$$\sin x = \frac{1}{2j}\left(e^{jx} - e^{-jx}\right), \tag{18}$$

$$\cos x = \frac{1}{2}\left(e^{jx} + e^{-jx}\right), \tag{19}$$

equation (17) was written as:

$$|W| = 2j \sin(k(x_2 - x_1)), \tag{20}$$

and assuming that the thickness of the plate $d = x_2 - x_1$, the equation was determined:

$$|W| = 2j \sin kd. \tag{21}$$

Determinant W_{A_1} results from equation (2.14) and is:

$$W_{A_1} = \begin{vmatrix} u_1 & e^{jkx_1} \\ u_2 & e^{jkx_2} \end{vmatrix}, \tag{22}$$

so:

$$|W_{A_1}| = u_1 e^{jkx_2} - u_2 e^{jkx_1}. \tag{23}$$

Using the record:

$$A_1 = \frac{|W_A|}{|W|}, \tag{24}$$

and substituting the expression (21), (23) into equation (24), A_1 was obtained:

$$A_1 = \frac{u_1 e^{jkx_2} - u_2 e^{jkx_1}}{2j \sin kd}. \tag{25}$$

Furthermore, the determinant W_{A_2} was calculated:

$$W_{A_2} = \begin{vmatrix} e^{-jkx_1} & u_1 \\ e^{-jkx_2} & u_2 \end{vmatrix}, \tag{26}$$

so:

$$|W_{A_2}| = u_2 e^{-jkx_1} - u_1 e^{-jkx_2}, \tag{27}$$

Analogously to (25) A_2 was determined:

$$A_2 = \frac{u_2 e^{-jkx_1} - u_1 e^{-jkx_2}}{2j \sin kd}. \tag{28}$$

Substituting (11) into equation (5) was written:

$$\sigma = c(-jkA_1 e^{-jkx} + jkA_2 e^{jkx}) - \frac{\varepsilon}{\varepsilon^S} D. \tag{29}$$

Using (10), the force was defined F_1:

$$-F_1 = -Ack(jA_1 e^{-jkx_1} - jA_2 e^{jkx_1}) - \frac{\varepsilon}{\varepsilon^S} D, \tag{30}$$

substituting determined A_1, A_2 to the equation (30), was obtained:

$$-F_1 = -Ack\left[j\left(\frac{u_1 e^{jkx_2} - u_2 e^{jkx_1}}{2j\sin kd} \right) e^{-jkx_1} - j\left(\frac{u_2 e^{-jkx_1} - u_1 e^{-jkx_2}}{2j\sin kd} \right) e^{jku_1} \right] - \frac{\varepsilon}{\varepsilon^S} D. \tag{31}$$

Carrying out the multiplication of expressions in brackets (31):

$$-F_1 = -Ack\left[\left(\frac{u_1 e^{jk(x_2-x_1)} - u_2 e^{jk(x_1-x_1)}}{2\sin kd} \right) - \left(\frac{u_2 e^{-jk(x_1-jx_1)} - u_1 e^{-jk(x_2-x_1)}}{2\sin kd} \right) \right] - \frac{\varepsilon}{\varepsilon^S} D, \tag{32}$$

and excluding the displacement u_1 and u_2 before the bracket:

$$-F_1 = -Ack\left[\frac{u_1\left(e^{jk(x_2-x_1)} + e^{-jk(x_2-x_1)} \right) - u_2\left(e^{jk(x_1-x_1)} + e^{-jk(x_1-x_1)} \right)}{2\sin kd} \right] - \frac{\varepsilon}{\varepsilon^S} D \tag{33}$$

And taking into account the dependence:

$$\cos kd = \frac{\left(e^{jk(x_2-x_1)} + e^{-jk(x_2-x_1)} \right)}{2}, \tag{34}$$

was obtained:

$$-F_1 = -Ack\left[\frac{u_1 \cos kd}{\sin kd} - \frac{u_2}{\sin kd} \right] - \frac{\varepsilon}{\varepsilon^S} D. \tag{35}$$

Finally, introducing trigonometric functions, equation (35) was written as:

$$F_1 = Ack\left[\frac{u_1}{\tan kd} - \frac{u_2}{\sin kd} \right] + \frac{\varepsilon}{\varepsilon^S} D, \tag{36}$$

In (36) the ralationship of force F_1, acting on the beginning of the system (Fig. 2), on the displacements u_1 and u_2 and its influence on the piezoelectric effect $\frac{\varepsilon}{\varepsilon^S} D$ were shown. Force F_2, determined from the equation (30) is equal to:

$$-F_2 = -Ack\left(jA_1 e^{-jkx_2} - jA_2 e^{jkx_2} \right) - \frac{\varepsilon}{\varepsilon^S} D, \tag{37}$$

and substituting the determinants A_1 (25), A_2 (28) to the equation (30), was obtained:

$$-F_2 = -Ack\left[j\left(\frac{u_1 e^{jkx_2} - u_2 e^{jkx_1}}{2j\sin kd} \right) e^{-jkx_2} - j\left(\frac{u_2 e^{-jkx_1} - u_1 e^{-jkx_2}}{2j\sin kd} \right) e^{jku_2} \right] - \frac{\varepsilon}{\varepsilon^S} D. \quad (38)$$

Carrying out the multiplication of expressions in brackets:

$$-F_2 = -Ack\left[\left(\frac{u_1 e^{jk(x_2 - x_2)} - u_2 e^{-jk(x_2 - x_1)}}{2\sin kd} \right) - \left(\frac{u_2 e^{jk(x_2 - jx_1)} - u_1 e^{-jk(x_2 - x_2)}}{2\sin kd} \right) \right] - \frac{\varepsilon}{\varepsilon^S} D, \quad (39)$$

and excluding displacements u_1 I u_2 before brackets:

$$-F_2 = -Ack\left[\frac{u_1\left(e^{jk(x_2 - x_2)} + e^{-jk(x_2 - x_2)} \right) - u_2\left(e^{-jk(x_2 - x_1)} + e^{jk(x_2 - x_1)} \right)}{2\sin kd} \right] - \frac{\varepsilon}{\varepsilon^S} D, \quad (40)$$

and using dependences (18, 19) written:

$$-F_2 = -Ack\left[\frac{u_1}{\sin kd} - \frac{u_2 \cos kd}{\sin kd} \right] - \frac{\varepsilon}{\varepsilon^S} D. \quad (41)$$

Finally, taking into account the trigonometric functions,

$$F_2 = Ack\left[\frac{u_1}{\sin kd} - \frac{u_2}{\tan kd} \right] + \frac{\varepsilon}{\varepsilon^S} D \quad (42)$$

was calculated.

Dependences of the forces and displacements acting on the system under consideration, taking into account the piezoelectric effect, are taking the following form:

$$\begin{cases} F_1 = \rho VA\left(\dfrac{u_1}{\tan kd} - \dfrac{u_2}{\sin kd} \right) + \dfrac{\varepsilon}{\varepsilon^S} D, \\ F_2 = \rho VA\left(\dfrac{u_1}{\sin kd} - \dfrac{u_2}{\tan kd} \right) + \dfrac{\varepsilon}{\varepsilon^S} D. \end{cases} \quad (43)$$

Analyzing the effects occurring in the piezoelectric plate, also the electrical parameters such as voltage on the plates of piezoelectric and current value were took into account. Voltage U is therefore expressed as a function of electric field:

$$U = \int_{u1}^{u2} E_p du, \tag{44}$$

where:

U - value of generated voltage on the linings of piezoelectric,

E_p - electric field intensity.

Integrating (44):

$$U = \frac{Dd}{\varepsilon^S} - \frac{\varepsilon}{\varepsilon^S}\left(u_2 - u_1\right), \tag{45}$$

where:

D- electric induction module, defined as:

$$D = \frac{i}{\omega A}, \tag{46}$$

finally the voltage in a function of current was shown as:

$$U = \frac{h}{\omega}\left(u_2 - u_1\right) + \frac{1}{\omega C_0} i, \tag{47}$$

where:

$$h = \frac{\varepsilon}{\varepsilon^S}. \tag{48}$$

Capacitance, depends directly on the dimensions of the plates, aswell as physicochemical properties, written in the form:

$$C_0 = \frac{\varepsilon^S A}{d}. \tag{49}$$

Introducing (46) into equations (36), (42), (46), dependences of replacement set of piezoelectric plate were received, plate characterized by three equations:

$$F_1 = Ack\left[\frac{u_1}{\tan kd} - \frac{u_2}{\sin kd}\right] + \frac{\varepsilon}{\varepsilon^S}\frac{i}{\omega A}, \tag{50}$$

$$F_2 = Ack\left[\frac{u_1}{\tan kd} - \frac{u_2}{\sin kd}\right] + \frac{\varepsilon}{\varepsilon^S}\frac{i}{\omega A}, \tag{51}$$

$$U = \frac{h}{\omega}\left(u_2 - u_1\right) + \frac{1}{\omega C_0} i \,. \tag{52}$$

Assuming signs :

$$k = \frac{\omega}{V}, \tag{53}$$

and

$$Z = \rho V A, \tag{54}$$

obtained the relations between the electrical and mechanical values of piezoelectric plates:

$$F_1 = Z\left[\frac{u_1}{\tan kd} - \frac{u_2}{\sin kd}\right] + \frac{hi}{\omega}, \tag{55}$$

$$F_2 = Z\left[\frac{u_1}{\tan kd} - \frac{u_2}{\sin kd}\right] + \frac{hi}{\omega}, \tag{56}$$

$$U = \frac{h}{\omega}(u_2 - u_1) + \frac{1}{\omega C_0} i \tag{57}$$

which are also written in a matrix form:

http://dx.doi.org/10.5772/54580

$$\begin{bmatrix} F_1 \\ U \\ F_2 \end{bmatrix} = \begin{bmatrix} \frac{Z}{\tan kd} & \frac{h}{\omega} & -\frac{Z}{\sin kd} \\ \frac{h}{\omega} & \frac{1}{\omega C_0} & -\frac{h}{\omega} \\ \frac{Z}{\sin kd} & \frac{h}{\omega} & -\frac{Z}{\tan kd} \end{bmatrix} \begin{bmatrix} u_1 \\ i \\ u_2 \end{bmatrix}. \tag{58}$$

4. Mapping matrix into graph

The values of susceptibility, admittance and characteristics were determined from the formula:

$$Y = \frac{1}{\det Z}\left((Z^M)^D\right)^T, \tag{59}$$

The matrix (58) was written as respond of system for operating extortion. The individual elements of matrix, presented as a flexibility, admittance and characteristics were recorded as follows: determined dependences of force from displacement are given in Table 1.

	F_1	U	F_2
u_1	$\frac{\left(\frac{h^2}{\omega^2}\right)-\left(\frac{\rho VA}{\omega C_0 \tan kd}\right)}{\det Z}$	-	$\frac{-\left(\frac{h^2}{\omega^2}\right)+\left(\frac{1}{\omega C_0}\frac{\rho VA}{\sin kd}\right)}{\det Z}$
i	-	-	-
u_2	$\frac{\left(\frac{h^2}{\omega^2}\right)-\left(\frac{1}{\omega C_0}\frac{\rho VA}{\sin kd}\right)}{\det Z}$	-	$\frac{-\left(\frac{h^2}{\omega^2}\right)+\left(\frac{\rho VA}{\omega C_0 \tan kd}\right)}{\det Z}$

Table 1. Dependences of force from displacement

The dependences between mechanical and electrical parameters were shown in Table 2.

In table 3 electrical dependences: voltage and current, so called admittance of piezoelectric system, were listed.

In order to determine graph, representing modeled system of piezoelectric plates, as the symbols in the matrix were used. The elements of matrix (58) assigned to the edges of graph are presented as:

	F_1	U	F_2
u_1	-	$\dfrac{\frac{h\rho VA}{\omega}\left(\frac{1}{\tan kd}-\frac{1}{\sin kd}\right)}{\det Z}$	-
i	$\dfrac{\frac{h\rho VA}{\omega}\left(\frac{1}{\tan kd}-\frac{1}{\sin kd}\right)}{\det Z}$	-	$\dfrac{\frac{h\rho VA}{\omega}\left(\frac{1}{\tan kd}-\frac{1}{\sin kd}\right)}{\det Z}$
u_2	-	$\dfrac{-\frac{h\rho VA}{\omega}\left(\frac{1}{\tan kd}+\frac{1}{\sin kd}\right)}{\det Z}$	-

Table 2. Dependences between mechanical and electrical parameters

	F_1	U	F_2
u_1	-	-	-
i	-	$\dfrac{-\left(\frac{\rho VA}{\tan kd}\right)^2+\left(\frac{\rho VA}{\sin kd}\right)^2}{\det Z}$	-
u_2	-	-	-

Table 3. Electrical dependences: voltage and current

$$Y_{11}=(x_1,F_1)=\frac{1}{\det Z}\left\{-\frac{\rho VA}{\omega C_0 tgkd}-\frac{h^2}{\omega^2}\right\} \tag{60}$$

$$Y_{12}=(x_1,U)=\frac{1}{\det Z}\left\{-\frac{\rho VA}{\sin kd}\frac{h}{\omega}-\frac{h}{\omega}\frac{\rho VA}{\sin kd}\right\} \tag{61}$$

$$Y_{13}=(x_1,F_2)=\frac{1}{\det Z}\left\{\frac{h^2}{\omega^2}-\frac{1}{\omega C_0}\frac{\rho VA}{\sin kd}\right\} \tag{62}$$

http://dx.doi.org/10.5772/54580

$$Y_{21} = (i, F_1) = \frac{1}{\det Z}\left\{\frac{h}{\omega}\frac{\rho VA}{tgkd} + \frac{\rho VA}{\sin kd}\frac{h}{\omega}\right\} \tag{63}$$

$$Y_{22} = (i, U) = \frac{1}{\det Z}\left\{-\left(\frac{\rho VA}{tgkd}\right)^2 + \left(\frac{\rho VA}{\sin kd}\right)^2\right\} \tag{64}$$

$$Y_{23} = (i, F_2) = \frac{1}{\det Z}\left\{-\frac{\rho VA}{\omega C_0 tgkd} - \frac{h^2}{\omega^2}\right\} \tag{65}$$

$$Y_{31} = (x_2, F_1) = \frac{1}{\det Z}\left\{\frac{h^2}{\omega^2} + \frac{\rho VA}{\omega C_0 \sin kd}\right\} \tag{66}$$

$$Y_{32} = (x_2, U) = \frac{1}{\det Z}\left\{\frac{h}{\omega}\frac{\rho VA}{tgkd} - \frac{\rho VA}{\sin kd}\frac{h}{\omega}\right\} \tag{67}$$

$$Y_{33} = (x_2, F_2) = \frac{1}{\det Z}\left\{\frac{h^2}{\omega^2}\frac{\rho VA}{tgkd} - \frac{\rho VA}{\omega C_0 tgkd}\right\} \tag{68}$$

The graphical representation of mapping is shown as:

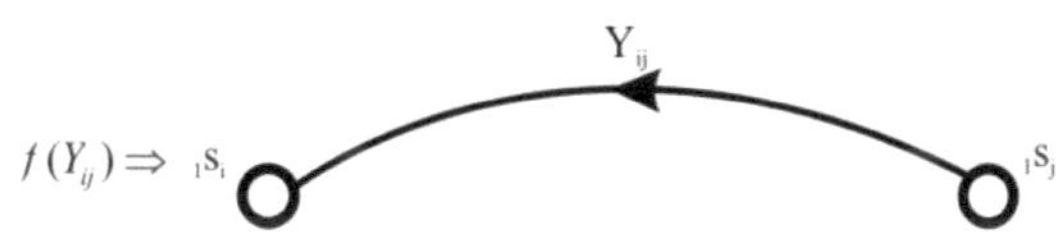

Figure 8. Mapping Y_{ij}

The symbol Y_{ij} is the mechanical flexibility, electrical admittance or characteristics of the system. In mapping of the parameters into the graph, mark Y_{ij} means the relationship between the vertex of graph, directed from the apex *i* to apex *j*, with the symbol *i*=*j*, then the following relationship were true:

$Y_{11}=Y_{10}$,

$$Y_{22} = Y_{20}, \tag{69}$$

$Y_{33}=Y_{30}$.

Dependences according to the index $j=0$ maps a connection of the vertex with the base vertex. Following this systematic, assignment by an edge of following relations was made:

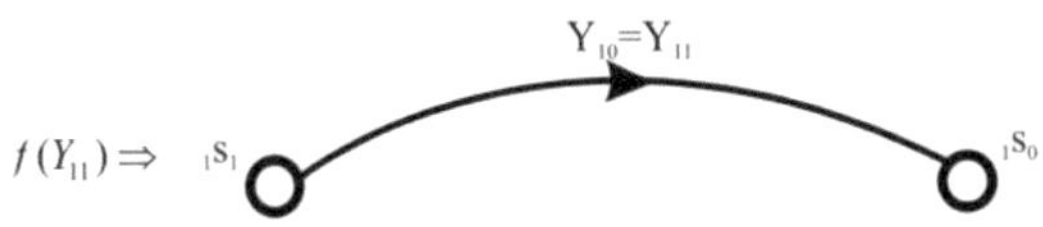

Figure 9. Mapping Y_{10}

where $f(Y_{11})$ is the mechanical flexibility;

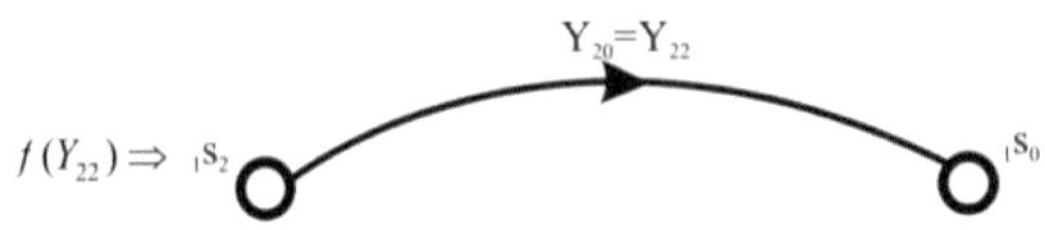

Figure 10. Mapping Y_{20}

where $f(Y_{22})$ is admittance of electrical system;

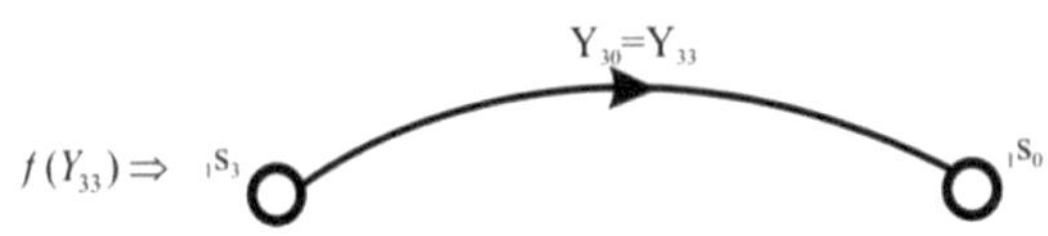

Figure 11. Mapping Y_{30}

where $f(Y_{33})$ is mechanical flexibility;

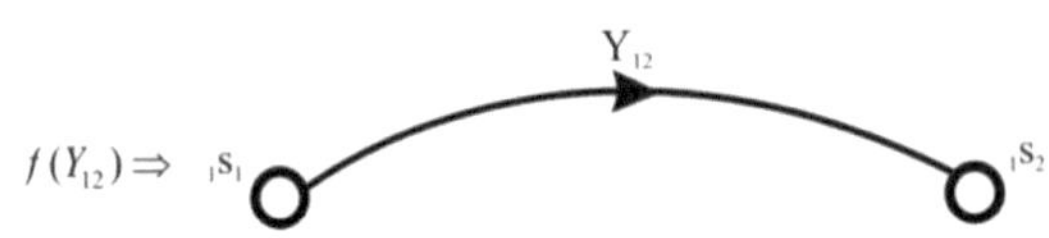

Figure 12. Mapping Y_{12}

where $f(Y_{12})$ is system characteristic;

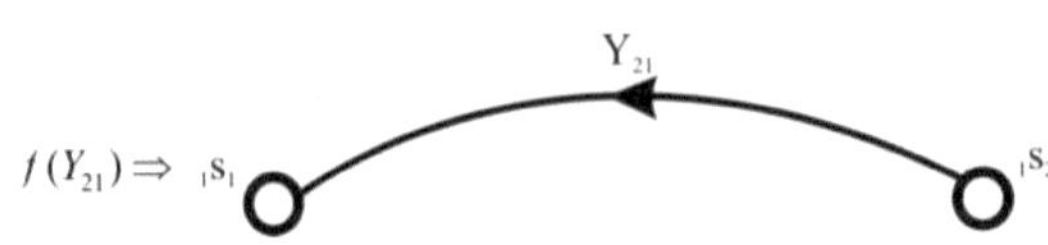

Figure 13. Mapping Y_{21}

where $f(Y_{21})$ is system characteristic;

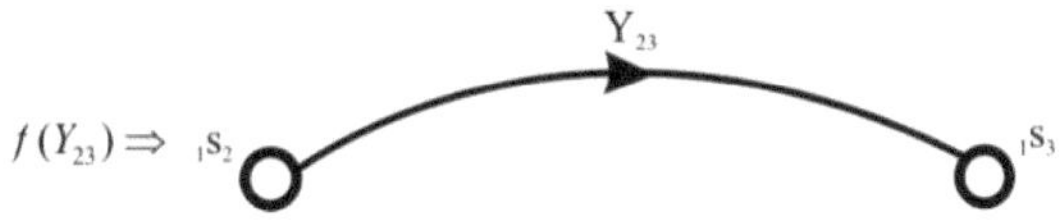

Figure 14. Mapping Y_{23}

where $f(Y_{23})$ is system characteristic;

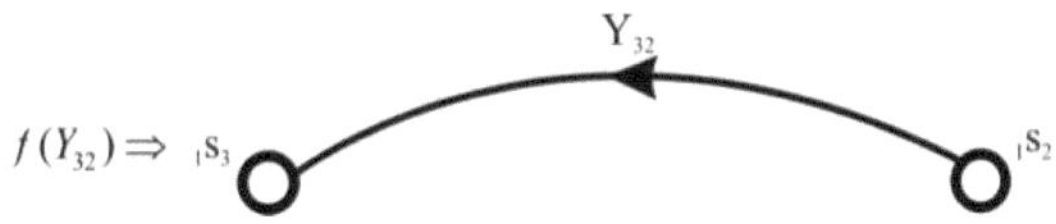

Figure 15. Mapping Y_{32}

where $f(Y_{32})$ is mechanical flexibility;

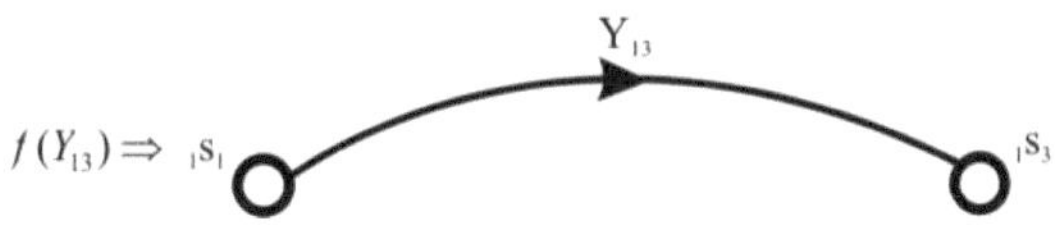

Figure 16. Mapping Y_{13}

where $f(Y_{13})$ is mechanical flexibility;

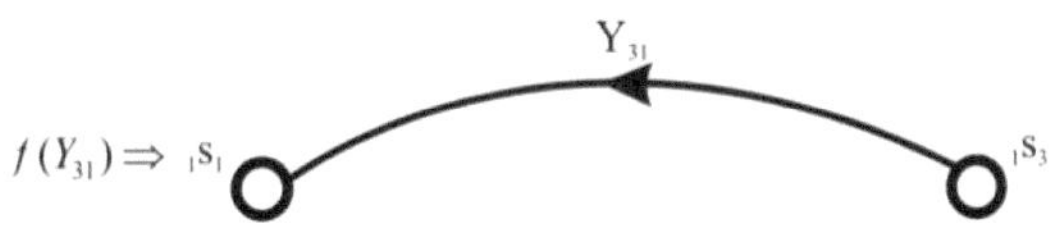

Figure 17. Mapping Y_{31}

where $f(Y_{31})$ is mechanical flexibility.

A set of drawings of the relation (fig. 9.) – (fig.17.), represents 4-vertex graph, were created and presented in Fig. 18.

$${}_2X = \{Y_{11}, Y_{22}, Y_{33}, Y_{12}, Y_{21}, Y_{31}, Y_{13}, Y_{23}, Y_{32}\} \tag{70}$$

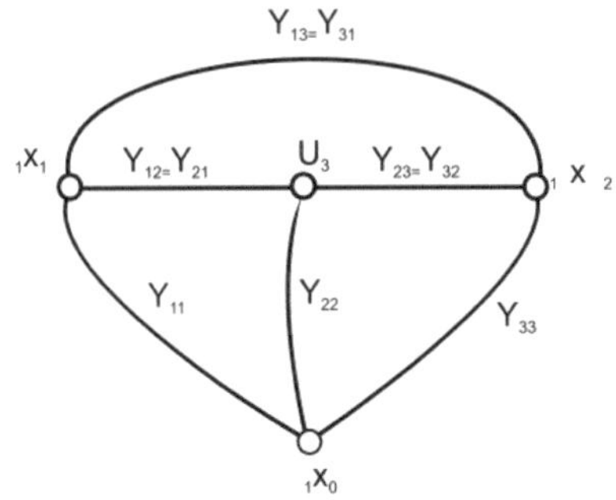

Figure 18. Geometric representation of mapping in the graph

In the rest of the work earlier created 4-vertex graph was replaced by structural number method to the 3-vertex graph.

5. Construction of the replacement graph

Furthermore, the use of an extended 4-vertex graph may prove to complicated calculations. In such case, a modelling of system using the replaced graph was performed. In order to maintain clearness of mapping, characteristics determined in paragraph 4 are indicated by Arabic numerals in parentheses, in accordance with (Bellert, 1981). As a consequence of introduction of the replaced graph, a graph presented in Fig. 19 was obtained. It is the basis for further network analysis methods.

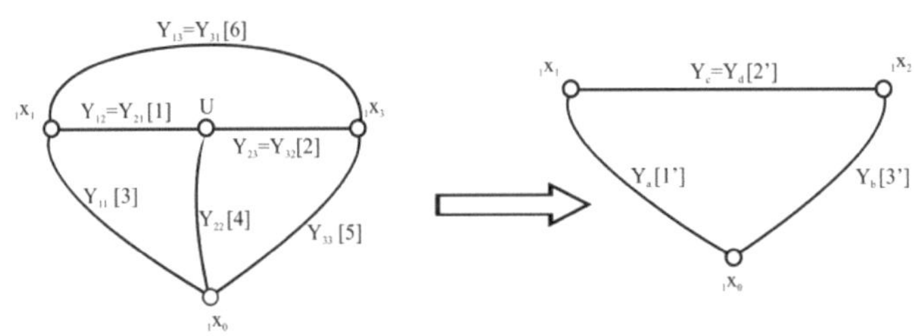

Figure 19. Construction of the replacement graph

As a result of insertion of replaced graph, replaced flexibility of the system was calculated by structural number method:

$$Y_b = Y_{3'} = \frac{\underset{Z}{\det}(A_{1'} \cap A_{2'})}{\underset{Z}{\det} A_{1'2'}} = \frac{Y_4(Y_2+Y_5)+Y_5(Y_1+Y_3)}{Y_1+Y_2+Y_4}. \tag{71}$$

$$Y_a = Y_{1'} = \frac{\underset{Z}{\det}(A_{2'} \cap A_{3'})}{\underset{Z}{\det} A_{2'3'}} = \frac{Y_1(Y_3 + Y_4) + Y_3(Y_2 + Y_4)}{Y_1 + Y_2 + Y_4}. \tag{72}$$

6. Chain equation of simple and complex system plate

On the figure 14 a piezoelectric plate with parameters distributed in the continuous way, the left and right end is free, was presented. The model of a single plate was marked by (i). Currently considered a model system is reduced system in the previous graph from 4-vertex to 3-vertex graph, as shown in Fig. 20.

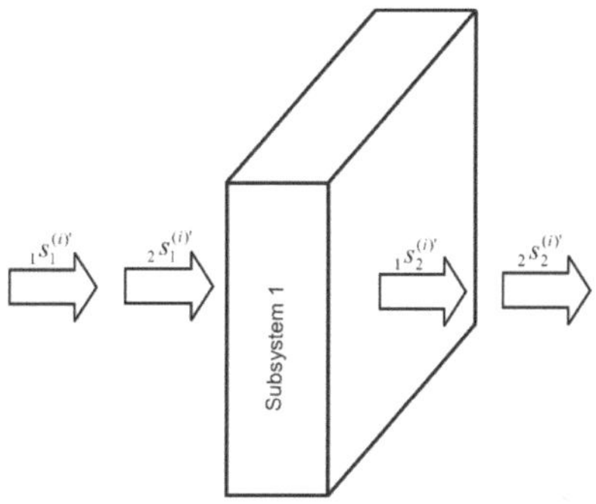

Figure 20. Model of single piezoelectric plate after reduction

Longitudinal vibrations of piezoelectric plate were considered, in the literature described also as thickness. The parameters specifying the system, in accordance with the previously accepted assumptions, were the sizes of input ${}_1s_1$, ${}_2s_1$ and output ${}_1s_2$, ${}_2s_2$ values, which were presented as:

$${}_1S^{(i)} = Y\,{}_2S^{(i)} \tag{73}$$

where:

Y is a value characterized input-output dependences.

The relations between displacements of plate, and the forces acting on them, written in matrix form:

$$\begin{bmatrix} {}_1s_1^{(i)} \\ {}_1s_2^{(i)} \end{bmatrix} = \begin{bmatrix} Y_a^{(i)} & Y_c^{(i)} \\ Y_d^{(i)} & Y_b^{(i)} \end{bmatrix} \begin{bmatrix} {}_2s_2^{(i)} \\ {}_2s_2^{(i)} \end{bmatrix} \tag{74}$$

Transforming the matrix (26) to the chain form expects to receive in the form of matrices:

$$\begin{bmatrix} {}_2s_1^{(i)} \\ {}_1s_1^{(i)} \end{bmatrix} = \begin{bmatrix} A_{11}^{(i)} & A_{12}^{(i)} \\ A_{21}^{(i)} & A_{22}^{(i)} \end{bmatrix} \begin{bmatrix} {}_2s_2^{(i)} \\ {}_1s_2^{(i)} \end{bmatrix} \tag{75}$$

where:

$$\begin{cases} A_{11}^{(i)} = \dfrac{Y_b^{(i)}}{Y_c^{(i)}}, \\ A_{12}^{(i)} = \dfrac{1}{Y_c^{(i)}}, \\ A_{21}^{(i)} = \dfrac{Y_c^{(i)}Y_d^{(i)} - Y_a^{(i)}Y_b^{(i)}}{Y_d^{(i)}}, \\ A_{22}^{(i)} = \dfrac{Y_a^{(i)}}{Y_d^{(i)}}. \end{cases} \tag{76}$$

In Fig. 21 the free system, consisting of two plates was presented. Superscript indicates the subsequent number of subsystem.

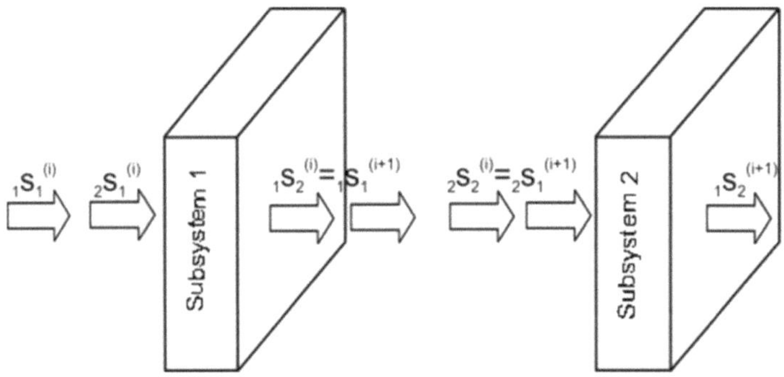

Figure 21. Diagram of a connection between the two cells and the relation between them

$$B = \begin{bmatrix} {}_2s_1^{(i)} \\ {}_1s_1^{(i)} \end{bmatrix} = \begin{bmatrix} A_{11}^{(i)} & A_{12}^{(i)} \\ A_{21}^{(i)} & A_{22}^{(i)} \end{bmatrix} \begin{bmatrix} A_{11}^{(i+1)} & A_{12}^{(i+1)} \\ A_{21}^{(i+1)} & A_{22}^{(i+1)} \end{bmatrix} \begin{bmatrix} {}_2s_2^{(i+1)} \\ {}_1s_2^{(i+1)} \end{bmatrix} \tag{77}$$

Finally, chain equation was written in general form:

$$A^{(k)} = A^{(i)}A^{(i+1)}. \tag{78}$$

After the operations carried out according to (79) it was found, that the chain matrix with cascade structure is the ratio of chain matrix of individual cells of the complex system. Obtained transition matrix is presented as:

$$B = \begin{bmatrix} {}_2s_1^{(i)} \\ {}_1s_1^{(i)} \end{bmatrix} = \begin{bmatrix} A_{11}^{(i+1)}A_{11}^{(i)} + A_{12}^{(i)}A_{21}^{(i+1)} & A_{11}^{(i)}A_{12}^{(i+1)} + A_{12}^{(i)}A_{22}^{(i+1)} \\ A_{21}^{(i)}A_{11}^{(i+1)} + A_{22}^{(i)}A_{21}^{(i+1)} & A_{21}^{(i)}A_{12}^{(i)} + A_{22}^{(i)}A_{22}^{(i+1)} \end{bmatrix} \begin{bmatrix} {}_2s_2^{(i+1)} \\ {}_1s_2^{(i+1)} \end{bmatrix} \tag{79}$$

Calculated coefficients (80) were substituted and the final form of the transition matrix was received:

$$B = \begin{bmatrix} B_{11}^{(k)} & B_{12}^{(k)} \\ B_{21}^{(k)} & B_{22}^{(k)} \end{bmatrix} \tag{80}$$

$$\begin{cases} B_{11}^{(k)} = \dfrac{Y_b^{(i)}Y_b^{(i+1)}}{Y_d^{(i)}Y_d^{(i+1)}} + \dfrac{-Y_a^{(i+1)}Y_b^{(i+1)} + Y_c^{(i+1)}Y_d^{(i+1)}}{Y_d^{(i)}Y_d^{(i+1)}}, \\ B_{12}^{(k)} = \dfrac{Y_b^{(i)}}{Y_d^{(i)}Y_d^{(i+1)}} + \dfrac{Y_a^{(i+1)}}{Y_c^{(i)}Y_d^{(i+1)}}, \\ B_{21}^{(k)} = \dfrac{(-Y_a^{(i)}Y_b^{(i)} + Y_c^{(i)}Y_d^{(i)})Y_b^{(i+1)}}{Y_d^{(i)}Y_d^{(i+1)}} + \dfrac{Y_a^{(i)}(-Y_a^{(i+1)}Y_b^{(i+1)} + Y_c^{(i+1)}Y_d^{(i+1)})}{Y_d^{(i)}Y_d^{(i+1)}}, \\ B_{22}^{(k)} = \dfrac{-Y_a^{(i)}Y_b^{(i)} + Y_c^{(i)}Y_d^{(i)}}{Y_d^{(i)}Y_d^{(i+1)}} + \dfrac{Y_a^{(i)}Y_a^{(i+1)}}{Y_d^{(i)}Y_d^{(i+1)}}, \end{cases} \tag{81}$$

In order to obtain the flexibility of the complex system, calculated coefficients of chain equation (79), was transformed to the basic form:

$$\begin{cases} Y_a^{(k)} = \dfrac{B_{22}{}^{(k)}}{B_{12}{}^{(k)}}, \\ Y_c^{(k)} = -\dfrac{B_{11}{}^{(k)}B_{22}{}^{(k)}}{B_{12}{}^{(k)}} - B_{21}{}^{(k)}, \\ Y_d^{(k)} = \dfrac{1}{B_{12}{}^{(k)}}, \\ Y_b^{(k)} = \dfrac{B_{11}{}^{(k)}}{B_{12}{}^{(k)}}. \end{cases} \tag{82}$$

Equation (83) is the components of the complex characteristics of the matrix taking into account obtained chain parameters of complex system.

7. Charts of simple and bimorph system

In this paragraph, graphical charts of characteristics of piezoelectric plates were shown. The parameters adopted for graphs plotting was presented in table 4.

No.	Symbol	Value	Unit
1	ρ	7.5	$[\frac{g}{cm^3}]$
2	$E=c_{33}$	150	$[GPa]$
3	A	3.1	$[cm^2]$
4	d	1	$[mm]$

Table 4. The parameters adopted for graphs plotting

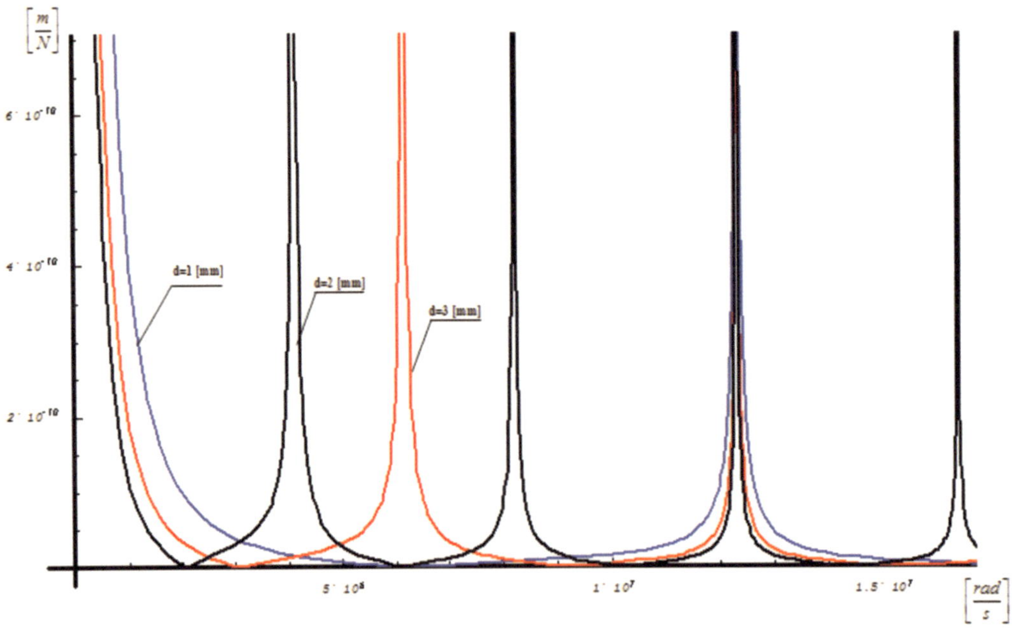

Figure 22. Characteristics of a single piezoelectric plate in the frequency domain, depending on the thickness of the plate

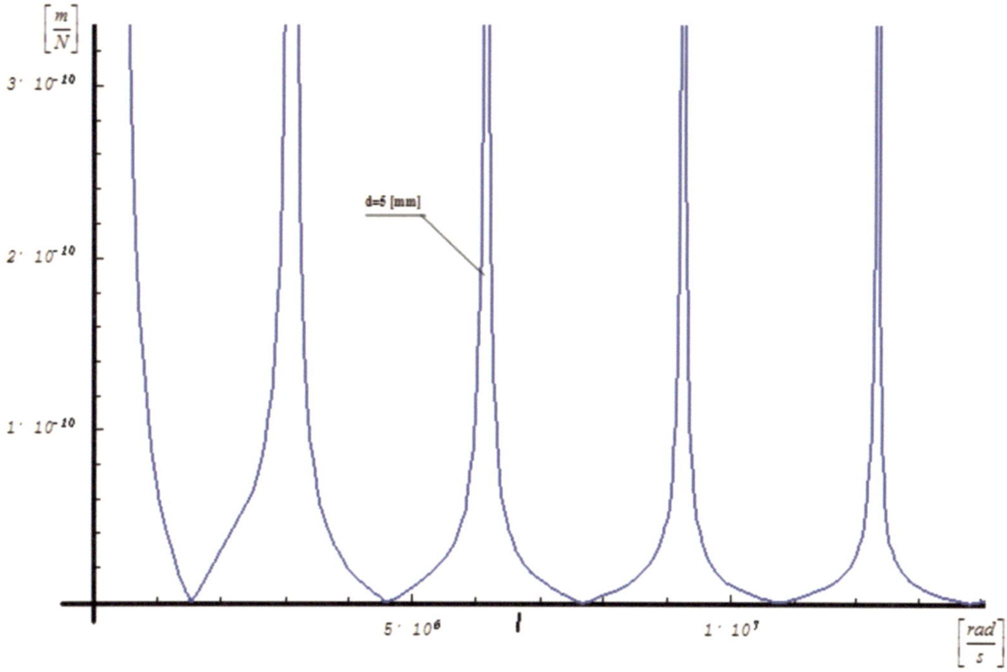

Figure 23. Characteristics of the combined plates of thickness 2and 3[mm] in frequency domain

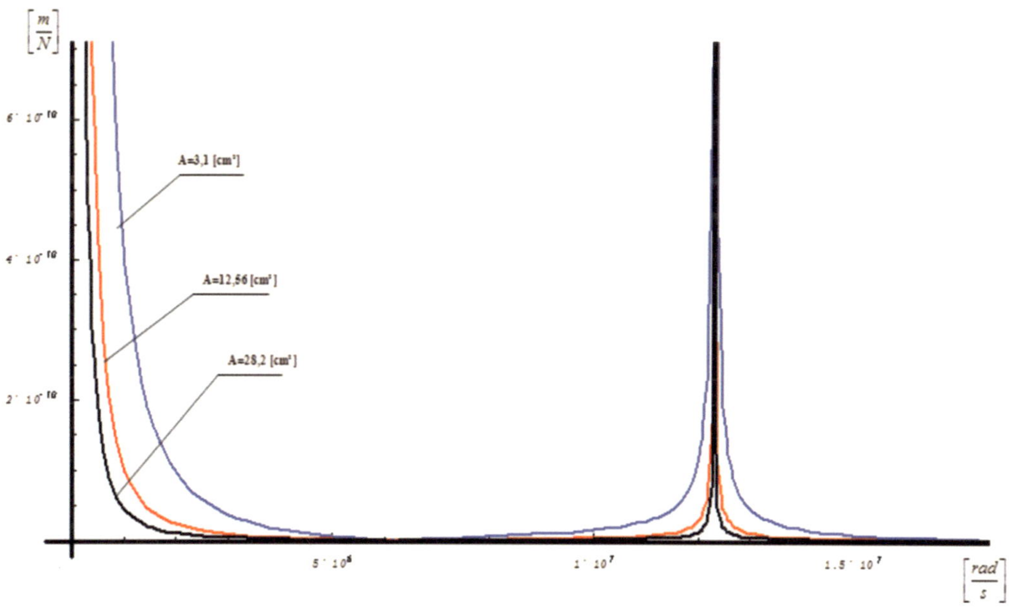

Figure 24. Characteristics of a single piezoelectric plate in the frequency domain, depending on the plate surface area

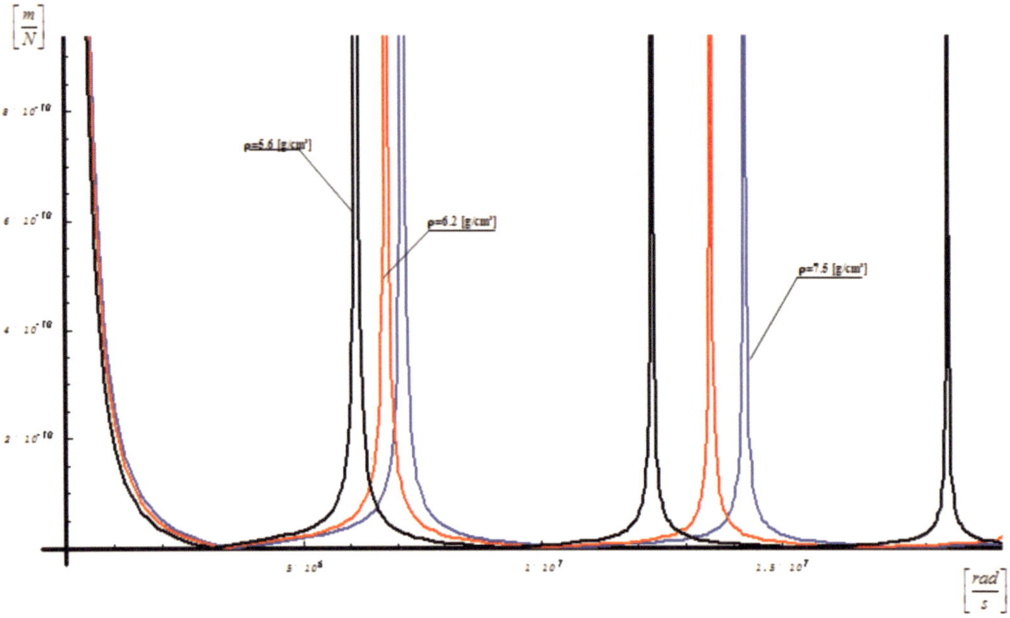

Figure 25. Characteristics of a single piezoelectric plate in the frequency domain, depending on the piezoelectric density

8. Conclusions

The chapter concerns the analysis of simple and complex piezoelectric systems, in order to determine the impact of piezoelectric plates parameters on the characteristics of the system. For a long time in the machine building are used subassemblies, whose operation is based on the piezoelectric phenomenon. In a researches of machine elements, on their surface piezoelectric sensors are glued, whereas to monitor the state plates are used transducers made from piezoelectric foil. Piezoelectric are often used in machine building also as assemblies, subassemblies or executive elements. Implementation of the piezoelectric system, which acts as a sensor or actuator is based on the selection of geometric dimensions of the plate, and their basic material parameters. In systems composed of several layers is also important piezoelectric plate number. Moreover, there are new problems at the design stage for the designers working in the field of machine building, concerning the application of both: single and stack plates. This matter is extremely important in terms of practical applications. For this reason it is necessary to conduct research whose main objective is to understand the phenomena associated with vibrations of complex piezoelectric systems.

Work is a continuation and development of decades researches at the Gliwice Center, consist in making the analysis of both the classical methods and non-classical. Take advantage of non-classical methods is a more general proposes from modeling in classical meaning. Resolves

http://dx.doi.org/10.5772/54580

simple and complex systems irrespective of the type and number of elements included in the test system.

Applied method of structural numbers method was presented and used previously in modeling mechanical systems.

Author details

Andrzej Buchacz and Andrzej Wróbel

Silesian University of Technology, Poland

References

[1] Arczewski, K. (1988). Structural Methods of the Complex Mechanical Systems Analysis, WPW, Warsaw

[2] Behrens, S, Fleming, A. J, & Moheimani, S. O. R. (2003). *Abroadband controller for shunt piezoelectric damping of structural vibration*. Smart Materials and Structures, , 12, 18-28.

[3] Bellert, S. (1981). *Chosen works,* PWN, 8-30100-247-6

[4] Bialas, K. (2012). *Mechanical and electrical elements in reduction of vibrations,* Journal of vibroengineering, 1392-8716, 14(1), 123-128.

[5] Bishop, R. E. D, Gladwell, G. M. L, & Michaelson, S. (1972). *Matrix analysis of vibration.* WNT, Warsaw

[6] Bolkowski, S. (1986). Theoretical electrical engineering. WNT, Warsaw

[7] Buchacz, A. (2004). *Hypergrphs and their subgraphs in modelling and investigation of robots.* Journal of materials processing technology. Complete, Elsevier, , 157-158, 3744.

[8] Buchacz, A, & Placzek, M. (2012). *The analysis of a composite beam with piezoelectric actuator based on the approximate method,* Journal of vibroengineering, 1392-8716, 14(1), 111-116.

[9] Buchacz, A, & Swider, J. (2000). Skeletons hypergraph in modeling, examination and position robot's manipulator and subassembly of machines. Silesian University of Technology Press, Gliwice.

[10] Ha, S. K. (2002). Analysis of a piezoelectric multimorph in extensional and flexular motions. Journal of Sound and Vibration, 253, 3, , 1001-1014.

[11] Kacprzyk, R, Motyl, E, Gajewski, J. B, & Pasternak, A. (1995). *Piezoelectric properties of nouniform electrets,* Journal of Electrostatics 35, , 161-166.

[12] Mason, W. P. (1948). Electromechanical Transducers and Wale Filters. Van Nostrand

[13] Sekala, A, & Swider, J. (2005). *Hybrid Graphs in Modelling and Analysis of Discrete-Continuous Mechanical Systems.* Journal of Materials Processing Technology, Complete Elsevier , 164-165, 1436-1443.

[14] Sherrit, S, Leary, S. P, & Dolgin, B. P. (1999). Comparison of the Mason and KLM equivalent circuits for piezoelectric resonators in thickness mode, Ultrasonics Symposium.

[15] Shin, H, Ahn, H, & Han, D. Y. (2005). *Modeling and analysis of multilayer piezoelectric transformer,* Materials chemistry and physics , 92, 616-620.

[16] Soluch, W. (1980). *The introduction to piezoelectronics*, WKiŁ, 8-32060-041-3

[17] Wróbel, A. (2012). *Model of piezoelectric including material damping,* Proceedings of 16th International Conference ModTech 2012, 2069-6736, 1061-1064.